Extraterrestrials

Extraterrestrials

A Philosophical Perspective

André Kukla

LEXINGTON BOOKS

A division of

ROWMAN & LITTLEFIELD PUBLISHERS, INC.
Lanham • New York • Toronto • Plymouth, UK

Published by Lexington Books
A division of Rowman & Littlefield Publishers, Inc.
A wholly owned subsidary of The Rowman & Littlefield Publishing Group, Inc.
4501 Forbes Boulevard, Suite 200, Lanham, Maryland 20706
http://www.lexingtonbooks.com

Estover Road
Plymouth PL6 7PY
United Kingdom

British Library Cataloguing in Publication Information Available

Library of Congress Cataloging-in-Publication Data

Kukla, André, 1942–
   Extraterrestrials : a philosophical perspective / André Kukla.
     p. cm.
   Includes bibliographical references (p. x) and index.
   ISBN 978-0-7391-4244-8 (cloth : alk. paper)
   ISBN 978-0-7391-4245-5 (pbk. : alk. paper)
   ISBN 978-0-7391-4246-2 (electronic)
   1. Extraterrestrial beings.   2. Science—Philosophy.
   I. Title.
   QB54.K85 2009
   576.8'39—dc22

                    2009032272

Printed in the United States of America

$\circledinfty^{\text{TM}}$ The paper used in this publication meets the minimum requirements of American National Standard for Information Sciences—Permanence of Paper for Printed Library Materials, ANSI/NISO Z39.48-1992.

Contents

Acknowledgments

Chapters 1 and 2 were published in a slightly different form in Studies in History and Philosophy of Science (2001, 32, 31–67). Chapter 4 was published in the British Journal for the Philosophy of Science (2008, 59, 73–88).

Chapter 1

On the Prospect of
an Extraterrestrial Encounter

My topic is extraterrestrial intelligence. Following current conventions, I use the abbreviation ETI to stand for three related concepts: (1) the abstract idea of extraterrestrial intelligence, (2) individuals who are both extraterrestrial and intelligent (as in "There's an ETI in the closet"), and (3) the hypothesis that there are ETIs. SETI is the search for ETIs, and CETI is the attempt to communicate with ETIs. In this chapter, I will try to answer the most basic question in extraterrestrial studies: what is the status of the ETI hypothesis? In the light of what we know, how likely is it that there are ETIs? We'll start with a brief history of the subject.

1.1   A BRIEF HISTORY OF ETI

Humans have been concerned with extraterrestrial questions more or less continuously since antiquity. Those who aren't familiar with the history of the subject may be surprised to learn that for the most part, it was the pro-ETI view that there *are* intelligent extraterrestrials that was in ascendancy. There were some dissenting voices—among them St. Augustine, St. Thomas Aquinas, and Albertus Magnus.[1] But they were greatly outnumbered by the pro-ETI forces, which included Aristotle, Plutarch, Lucretius, Nicholas of Cusa, Giordano Bruno, Gassendi, Locke, Lambert, Kant, Thomas Wilkins, Christian Huygens, and Bernard de Fontenelle.[2]

The two ancient writers who dealt most extensively with the topic of ETI were Plutarch and Lucretius. Both of them were pro-ETI, but each gave different reasons for his position. Their arguments presaged the two basic strategies for arguing in favor of ETI for centuries to come. Some features of their arguments are parts of the pro-ETI armamentarium to this day. Plutarch inferred that intelligent life exists on the moon from four premises: (1) the earth has no privileged position in the universe; (2) the arrangement of the heavens indicates intelligent design; (3) the moon is sufficiently like the earth to support life; (4) if the moon didn't have life, it would exist to no purpose, and this conflicts with the assumption of intelligent design. Of course, we know now that the conclusion of this argument is false. There's no life on the moon—we've been there. But the same teleological argument was made by Bruno and Nicholas of Cusa for the billions of stars: would an intelligent designer have wasted so much space?

This is not the place to argue about the legitimacy of teleological explanations. Grant that the universe has been intelligently designed. Doesn't the empirical refutation of Plutarch's original lunar hypothesis show that the general argument-form is invalid, even when it's applied to galaxies? Apparently, the Intelligent Designer *is* willing to build entire worlds without life. There must be something about Her plan that we don't understand. This brings up the question of Plutarch's third premise—that the moon is sufficiently like the earth to support life. The occurrence of this premise in the argument shows that Plutarch didn't suppose that an intelligently designed universe would not contain uninhabited planets. If that were his view, then the third premise would be superfluous. Evidently, what Plutarch's teleological intuition told him was that an intelligent designer wouldn't make *inhabitable* planets and then leave them uninhabited. Let's grant this hypothesis about intelligent design as well. Then the empirically established fact that the moon contains no life doesn't by itself invalidate the later galactic version of the argument—it just shows the premise about the moon to be false. But also, it shows that the bare existence of billions of stars doesn't by itself make the teleological argument work. It's not enough to establish that there's sufficient *space* for other life forms. The Apollo missions to the moon already tell us that the (presumed) intelligent design of the universe must be compatible with the existence of large, uninhabited bodies. The Plutarchian argument needs the premise that among the billions of stars, there are other *inhabitable* places. There was nothing about Cusa's or Bruno's bare observation of the teeming stars in the night sky that could have provided them with that premise. Contemporary proponents of SETI think that they have good reasons for supposing that there are many other inhabitable places in the universe. To be sure, they don't embed this hypothesis in a teleological argument. But both the Plutarchian argument and the contemporary SETI arguments for the existence of ETIs rely on the hypothesis that there are other inhabitable places. We'll examine the contemporary arguments in due time. If the case for their inhabitability hypothesis should prove to be compelling, it would lend a measure of credence to the Plutarchian argument as well. But the galacticized version of Plutarch's argument by itself is inadequate, even if we accept its teleological assumptions. Maybe the Intelligent Designer made the stars for purely decorative reasons.

There's also a more parochial class of pro-ETI argument from deistic premises. The hypothesis of a plurality of worlds inhabited by intelligent beings poses no special problems for devout Hindus or Buddhists. But it's a central idea of the Judeo-Christian tradition that there's a unique relationship between God and human beings. The Judeo-Christian God spends an enormous amount of time concerning Himself [sic] with human affairs. This observation doesn't by itself entail that Judeo-Christian orthodoxy is incompatible with ETI—for it's possible that God is simultaneously carrying out similar activities throughout the universe—that He processes the needs of His myriads of cosmic children in parallel mode. But it's of the essence of Christianity that God sacrificed His *only* begotten son to save humanity. If the only begotten son of God merely saves *some* of the intelligent beings in the universe, what has been the disposition of the rest? As Thomas Paine noted, the existence of a plurality of worlds renders Christianity "little and ridiculous, and scatters it in the mind like feathers in the air" (Paine, 1948, p. 44). What follows from this observation? As the saying goes, one person's *reductio ad absurdum* is another's discovery of a novel consequence. That a plurality of worlds renders Christianity little and ridiculous can be regarded as an argument against the plurality of worlds or as a proof of the smallness and silliness of Christianity.

Lucretius' argument is much more modern than Plutarch's. For one thing, he dispenses entirely with teleological assumptions. According to Lucretius, the conclusion of ETI follows from purely mechanical assumptions, together with the premise that the universe is infinite in extent:

> Granted, then, that empty space extends without limit in every direction and that seeds innumerable in number are rushing on countless courses through an unfathomable universe under the impulse of perpetual motion, it is in the highest degree unlikely that this earth and sky is the only one to have been created. . . . This follows from the fact that our world has been made by nature through the spontaneous and casual collision and the multifarious, accidental, random and purposeless congregation and coalescence of atoms whose suddenly formed combinations could serve on each occasion as the starting-point of substantial fabrics—earth and sea and sky and the races of living creatures. (Lucretius, 1951, pp. 91–92).

What replaces design here is the same thing that Darwin replaced it with in his account of the origin of species: chance. The argument is that if the world is an infinite collection of atoms jostling about at random, then any conglomeration whatever is going to occur many times over. The same argument-form is employed in drawing the conclusion that a roomful monkeys hitting typewriter keys at random will eventually produce the verbatim text of *Hamlet*—and that if we wait long enough, they'll produce it again. This type of mechanical argument is not merely still around—somewhat amended versions of it are presently the mainstay of the pro-SETI case. We'll see that in some ways, Lucretius' original argument is more robust than some of its descendants. The vicissitudes of this argument are part of the modern, scientific story and will be dealt with at a later time. In any case, whether from teleological or mechanistic assumptions, the existence of ETIs was the received view by 17th and 18th centuries, and has remained so, though there have been notable dissenting voices.

Plutarch's and Lucretius' arguments bear on the substantive question whether there are any ETIs. History also records the voices of some who objected, not to the ETI hypothesis, but to the idea that the ETI hypothesis raises an interesting or important question. Montaigne, Milton, and Pope held the view that speculation about ETIs is a waste of time, and that we would do better to think about more mundane issues. Let's call the question whether the ETI hypothesis is worth pursuing the *metaquestion* of extraterrestrial studies. I'll deal with the substantive question in this chapter and with the metaquestion in chapter 2. This may seem like putting the cart before the horse—for shouldn't we ascertain that the substantive question is worth an answer before we try to answer it? It would be nice always to know whether a project is pursuitworthy before we invest an iota of time or energy in its pursuit. But this ideal is unrealizable. In fact, it leads to an infinite regress—for wouldn't we have to know that the *meta*question (i.e., whether the ETI hypothesis is worth pursuing) is itself pursuitworthy before devoting an iota of effort to it? In practice, there's no way to avoid an initial speculative investment of effort. So we might just as well speculate that the ETI question is pursuitworthy as that the metaquestion is pursuitworthy.[3]

Moreover, the resolutions of both questions are inextricably intertwined. If the discovery of ETIs promises to bring enormous benefits in its train, that would be a reason to devote large quantities of resources to looking for them. But this reason would be negated if the probability of there *being* any ETIs to contact were vanishingly small. So the answer to the metaquestion depends in part on the answer to the substantive question. The most reasonable way to proceed is, I think, clear and uncontentious. Before devoting millions more to the search for ETIs, we need a low-cost, paper-and-pencil initial assessment of the prospects for the success of such a project. It's true that to do even this much is to presume that the pursuit of extraterrestrial issues has some value. But the working time of scholars is cheaply bought. Surely, everyone will agree that the pursuit of the ETI hypothesis is worth at least a few weeks of a few scholars' time, plus stationery—say twenty thousand 1999 US dollars. If the result of this cheap initial assessment of the ETI hypothesis is that there's no chance of our ever finding out that there are ETIs, then that may be the end of the story. If the result of the initial assessment is anything else, then it becomes one factor among others in the determination of whether larger-scale investigations of the ETI question are a good investment.

1.2   INDUCTIVE ARGUMENTS FOR ETI

How plausible is it to suppose that there are ETIs? Discussions of this issue are generally guided by the frequently-cited Drake equation.[4] There are several minor variants of this equation in the ETI literature. Mash gives the following version:

$$N = N_* f_p n_e f_l f_i f_c f_L,$$

where

N = the number of presently existing advanced civilizations in the galaxy (where "advanced" means "at least capable of and engaged in interstellar communication")

N_* = the number of stars in the galaxy

f_p = the fractions of those stars with planets

n_e = the average number of planets in the solar system whose environments are suitable for life

f_l = the fraction of habitable planets on which life occurs

f_i = the fraction of life-bearing planets on which intelligence evolves

f_c = the fraction of intelligence-bearing planets on which advanced civilizations arise

f_L = the fraction of advanced civilizations existing at this time (as determined by the average lifetime of a civilization divided by the present age of the galaxy, or, more precisely, the age of the civilization-bearing galaxy; Mash, 1993, pp. 210–211).

The last two factors, f_c and f_L, are interestingly different from the rest. As is noted in the definition of N, "advanced" civilizations are those that have the technological capability for sending and receiving interstellar signals. Of course, everyone concedes that it's possible to be intelligent without possessing this capability. We didn't possess it ourselves until

quite recently. So why the restriction? The reason, as Mash notes, is that
the Drake equation is

> oriented toward the funding of SETI projects, that is, projects that intend to
> *search* for extraterrestrial civilizations, rather than toward the more academic
> question of whether other civilizations might exist anywhere at any time
> (p. 211).

So, we need to distinguish two substantive hypotheses: the *ETI hypothesis*
that there are ETIs, and the *SETI hypothesis* that we can *make contact with*
ETIs. It's conceivable that a general argument based on firmly established
astrophysical, biological, and psychosocial principles might convince us that
it's overwhelmingly likely that ETIs exist, but that it's also overwhelmingly
*un*likely that we will ever be able to make contact with them. If this were to
be the case, it would make no theoretical or practical difference whether some
of these ETIs happen to exist contemporaneously with our own civilization.
The bare fact of their existence, anywhere and at any time, would make no
less of an impact on our lives and thoughts than the fact of their existence
now. Thus it's more perspicuous to equate the ETI hypothesis with the general
proposition that there are ETIs somewhere and at some time. This means that
the Drake equation relevant to the ETI hypothesis is at least two terms shorter
than the equation relevant to the SETI hypothesis. Let N' be the number of
planets in the galaxy harboring intelligent life at any time. Then

$$N' = N_* f_p n_e f_l f_i.$$

When it comes to SETI, however, even the full equation for N given above
is two terms short. It's not enough to know that there exist temporally overlapping extraterrestrial civilizations engaged in interstellar communication.
Conceivably, we might come to be persuaded that this is so on the basis of
a theoretical argument. But this would not yet be SETI success. To succeed
at SETI, we have to *detect their signals.* Moreover, having detected their
signals, we need to ascertain that these signals are *expressions in a language*
rather than a peculiar, inanimate phenomenon, or the radio equivalent of a
bird song. The most straightforward way to identify a series of signals as
linguistic is to *decode* it—to figure out a scheme for translating the signals
into our language and to test our scheme on a new series of signals. Donald
Davidson (1974) has famously expressed the view that decoding is the *only*
way to ascertain that a series of signals belongs to a language. I will assess the
merits and demerits of this view in chapter 3. But it's worth noting here that
if Davidson is wrong, then *CETI* (not SETI) success requires a third extra-
Drakean step beyond detecting extraterrestrial signals and ascertaining that
they're linguistic: it requires also that these linguistic signals be decoded. If,

on the other hand, Davidson is right, then the second step is identical to the third, and SETI and CETI succeed or fail together.

As would be expected, the Drake equation incorporates factors from all the major branches of science—the physical, the biological, and the psychosocial. Most of the discussion among SETI researchers has centered around the physical and biological factors, represented at the beginning and middle of the equation. The psychosocial factors at the end have not usually been the loci of sustained intellectual dispute. What accounts for this disparity? I attribute it to the fact that SETI funding has gone almost exclusively to physical and biological scientists.[5] An alternative explanation is that there isn't as much to be said about the psychosocial issues. There are no canonical psychosocial theories that can be brought to bear on the issues. Physics and biology, on the other hand, possess the heavy intellectual artillery that must surely make a difference in the field. This sounds very plausible. However, more than one critic has noted that our physical and biological theories, excellent as they are in their own right, fail to make contact with the task of estimating the Drake factors (McMullen, 1980; Baird, 1987; Mash, 1993). The paths from current physical and biological theory to estimates of the Drake factors are all paved with adhockery. In the end, the physical and biological parameters of the Drake equation are in the same epistemic boat as the psychosocial: our estimates of them are not dictated by theoretical principles.

McMullen draws a harsh lesson from this state of affairs:

> Until we have a theory of some sort, we cannot attach a theoretical probability of any kind to an outcome . . . there is no responsible way of [estimating the Drake factors] except in the light of a detailed and properly warranted theory (McMullen, 1980, p. 84).

McMullen's judgment may be too severe, for the likelihood of a hypothesis can also be estimated on the basis of brute enumerative induction, even in the absence of a detailed relevant theory. If numerous, randomly chosen samples of the raven population turn out to be composed of about 60 percent black ravens and 40 percent non-black ravens, it would be rational for us to ascribe a probability of .6 to the hypothesis that the next raven we encounter will be black—even if we lack a theoretical account of why the observed frequencies are what they are.

SETI sceptics point out, however, that this possibility will not help us to develop rationally grounded opinions about the Drake factors. The reason is simple: you can't base an inductive inference on a single case. If we had observed only a single raven and found it to be black, we would have no more inductive warrant for expecting future ravens to be black than for expecting that other ravens will have precisely the same scratch on their left foot as one that was found on the original, observed raven. With just a single case before us, there's no basis for making an educated guess as to which of the observed properties are connected in some law-like way to ravenhood and which are merely accidents. The fact that we've seen one inhabitable planet spawn life doesn't give us inductive grounds for believing that any of the other inhabitable planets in the galaxy will also spawn life. Nor is a single case sufficient for making a probabilistic estimate. In the absence of relevant theory, the only way to make such an estimate is to equate the probability of the target event to the frequency of target events in the total sample of observed events. If we applied this procedure to the Drake factors, we would conclude that the probability that a randomly chosen star has a planet which harbors an advanced civilization is *one*, since the observed frequency of such stars in the set of adequately examined stars is one over one. The catch, of course, is that the reliability of the probabilistic estimate increases with the size of the observed sample. When the sample contains a single element, the estimate is maximally unreliable.

So the single case of intelligent life that we're familiar with doesn't underwrite any estimate for the probability that there are other cases. However, it's widely believed in SETI circles that (1) the single case does allow us to conclude that the probability of there being intelligent life on another, randomly chosen world cannot be exactly zero, and that (2) this is ultimately enough of a concession to warrant acceptance of the hypothesis that there *is* intelligent life on *some* other worlds. Let's look at both parts of this package. The rationale for the first part runs as follows: the single case of intelligent life on earth establishes that the evolution of intelligent beings is not forbidden by the laws of nature, and what is not forbidden has a non-zero probability of recurring. There are hidden complexities in this inferential step that have never been adequately analyzed. The main problem with it is that the occurrence of an event doesn't logically entail that the laws that produced it must allow for a second production. It's easy to construct logically possible universes whose laws entail that certain events are unique. For example, consider a universe that begins to exist at time t_0, containing only a single particle of size S. Suppose also that the only law of nature in this universe is that after a given time interval every particle in the universe emits an "offspring" particle that's half the size of the original. Thus, after a single time interval, there are two particles: the original parent particle of size S and an offspring of size S/2. After the second time interval, there are four particles: one of size S, two of size S/2, and one of size S/4, and so on. This is a well-behaved, logically possible universe. But in this universe, it would be a mistake to suppose that just because there exists one instance of a particular type of item, there must be a non-zero probability that there are other instances of the same item. In this universe, there exists a particle of size S—namely, the original parent particle. Nevertheless, the probability of encountering *another* particle of size S is exactly zero. Additional items of size S have the same status in this universe as perpetual motion machines have in ours: they're ruled out by natural law. The lesson is that the assumption that the universe is a lawful place doesn't by itself underwrite the conclusion that what happens once must be allowed to happen again. It must be regarded as an open question whether the laws of *our* universe entail this conclusion.

Moreover, the inference from our single case to the non-zero probability of ETI wouldn't do ETI advocates any good even if they were permitted to make it. There are numerous arguments in the ETI literature that attempt to parlay the non-zeroness of the probability of ETI into a reason for accepting ETI as a plausible hypothesis. All of these arguments, however, commit one or both of two errors: (1) the error of one-sided estimation, and (2) the error of over-precision. The main example of the first error is to be found in the argument that the universe is so vast that even extremely unlikely events are likely to occur at some place or other:

> [T]he idea that *we* are the only intelligent creatures in a cosmos of a hundred million galaxies is so preposterous that there are very few astronomers today who would take it seriously (Clarke, 1973, p. 90).
>
> With the Universe constructed on so vast a scale, it would seem inherently improbable that our small Earth can be the only home of life (Jones, 1940, p. 7).
>
> In fact, there are billions of galaxies, each containing billions of stars. This is the single most important reason for optimism in the search for life. In a universe so vast, with stars as numerous as grains of sand, it is hard to imagine that the conditions for life have not arisen elsewhere (McDonough, 1987, p. 59).

To call this an argument is to insult arguments. In fact, it's nothing more than a profession of one's faith that the number of stars is large enough to overwhelm the smallness of the probability that there are ETIs associated with a randomly selected star. Certainly, this conclusion doesn't follow from the admission that the probability of ETI is non-zero. For however many stars there may be, so long as the number is finite there's going to be a non-zero probability for intelligent life that's so ultramicroscopically small that the net probability of ETI anywhere in the universe will still be as close to zero as to make no difference. For example, the estimated number of stars in the observable universe is 10^{22}. Let's suppose that this is a drastic underestimation. Let's suppose that there are actually 10^{100} stars. Let's also suppose that all the Drake factors except one are extremely favorable for the SETI hypothesis: that virtually all stars have planets, that the average number of planets per solar system whose environment is suitable for life is 100 (we live in an extremely deprived system), that all planets suitable for life actually develop life, that advanced civilizations arise every place where intelligence develops, and that once an advanced civilization arises it lasts forever. The only fly in the ointment is that the probability that intelligence develops where life exists is only 10^{-150}. Then, even though there are 10^{102} planets harboring life, the probability that any of them contains *intelligent* life is only 10^{-48}. If that isn't negligible enough to eliminate ETI from serious consideration, we can run through the same example with a probability of 10^{-1000} for the development of intelligence, or $10^{-1,000,000}$. The point is that the thesis that the probability of ETI around a given star is non-zero is compatible with any of these estimates.

That's all that needs to be said about the raw appeal to large numbers. But Mash (1993) has an exercise that helps to dissolve the unwarranted intuition that the immensity of the universe must overwhelm the minuteness of the probability of ETI in any given locale. The probability of passing the Drake gateways gets smaller and smaller as more of them are taken into account: it's likelier that there are inhabitable planets than that there are inhabitable planets that actually possess life; it's likelier that there's extraterrestrial life than that there's extraterrestrial *intelligent* life; and so on. The Drake equation stops at intelligent beings who are sufficiently like us that they engage in their own SETI project. The pro-SETI intuition is that, though the probability gets smaller as we traverse the various Drake factors, we're left at the end with a probability that's large enough to be swamped by the vastness of the universe. Now what happens if we add an additional requirement to the end of the Drake equation—say that the ETIs not only engage in interstellar communication but that they also speak English? To be sure, the resulting scenario would be orders of magnitude less likely than the scenario that merely specifies that the ETIs engage in interstellar communication. But why not say, once again, that the vastness of the universe ensures that even these minute probabilities are actualized? After all, it would be utterly arbitrary and unreasonable to suppose that net probabilities cease to be overwhelmed by the vastness of the universe at the precise point where the conventional Drake equation leaves off—that it's plausible that there are ETIs who build radio telescopes somewhere in the universe, but that it isn't plausible that there are English-speaking ETIs who build radio telescopes.

And why stop at speaking English? Why shouldn't we suppose that the vastness of the universe overwhelms the extreme unlikelihood of there being ETIs who have come upon Velcro fasteners, and whose teenagers amuse themselves with skate boards, hula-hoops, and yo-yos? In fact, why doesn't the vastness of the universe make it plausible that there's a twin earth where someone just like you reads these very words? The argument is the same: it's scientifically possible for there to be a twin earth—or so we have assumed; therefore, given the billions upon billions of stars that each provide an occasion for it, it's likely to have happened somewhere. But surely there's some point along the chain of diminishing probabilities where even the most ardent SETI enthusiast will balk—if not at the existence of one twin earth, then at the existence of ten, or a million of them. However large the universe may be (so long as it's finite), it isn't going to be vast enough to accommodate any and all scientifically possible scenarios. There has to be a cut-off point along the series of diminishing probabilities beyond which the vastness argument no longer applies. But then, what reason do we have for supposing that intelligent life lies on one side of this cut-off point, while the employment of Velcro fasteners and skate boards lies on the other? Rescher has put the matter very succinctly:

> Admittedly, cosmic locales are very numerous. But probabilities can get to be very small: no matter how massive N, there is that diminutive 1/N that can countervail against it (Rescher, 1985, p. 109),

It's important to realize that this argument cuts both ways. Mash reminds us that

> "the corollary" to [Rescher's assertion] is that no matter how small a fraction 1/N faces us, there is always that big number N that can squash it (Mash, 1993, p. 220).

SETI sceptics have tended to make the symmetric error of arguing one-sidedly for the rejection of ETI on the grounds that 1/N is very, very small. An example is Mayr's "evolutionary" argument:

> The point I am making is the incredible improbability of genuine intelligence emerging [from an evolutionary process]. There were probably more than a billion species of animals on earth, belonging to many millions of separate phyletic lines, all living on this planet earth which is hospitable to intelligence, and yet only a single one of them succeeded in producing intelligence (Mayr, 1985, p. 28).

At first glance, it looks as though this anti-SETI evolutionary argument is superior to the pro-SETI vastness-of-the-universe argument we've just considered: the latter employs an inductive inference based on a single case, whereas the former's inductive inference is based on billions of cases. Of course, the billions of cases aren't randomly selected. They're all species that evolved on the same planet. Mayr characterizes the earth as a planet "which is hospitable to intelligence." But this is nothing more than a guess. It could turn out that there's something about the terrestrial environment that *inhibits* the evolution of intelligence, as compared to other planetary environments which

are suitable for life. Alternatively, there might be something about evolutionary processes generally that militates against the appearance of more than one
intelligent species per planet. (Perhaps intelligence is inevitably accompanied
by a xenophobia so intense that the first intelligent species to appear extermi-
nates all the near-intelligent competitors.)

An even more telling critique of the evolutionary argument is that it's an
induction based on a single case after all. The one-in-billions figure alluded
to by Mayr is the observed frequency of intelligent species among total species. But this frequency does not provide an estimate of any of the factors in
the Drake equation. The factor that comes closest to it is f_i, the fraction of
life-bearing planets on which intelligence evolves. But the appropriate ratio
for estimating this factor is the observed incidences of intelligent species
among total *evolutionary histories*. It makes no difference to f_i if evolutionary processes profligately produce myriads of unintelligent species. All that
matters is how frequently evolution produces intelligence. But, of course, we
have only one evolutionary history upon which to base our opinion. And *that*
observed frequency is one over one.

But let's suppose that the contemplation of the twists and turns of evolutionary history on earth persuades us all that the probability that intelligence
evolves on a life-bearing planet is extraordinarily small. The question once
again arises: is it small enough? The anti-SETI biologists seem to be guilty
of relying on an unreasoned faith which is complementary to the unreasoned
faith of the pro-SETI astronomers. In the conflict between enormous Ns and
minute 1/Ns, the astronomers arbitrarily take the side of the Ns and the biologists arbitrarily align themselves with the 1/Ns. One-sided appeals to enormity
or minuteness are not enough. To ground an opinion about SETI, we need to
make absolute estimates of *all* the Drake factors, multiply them together, and
see what we get. Such estimates are available in the SETI literature (Dole,
1964; Sagan, 1980). But in the absence of both a theoretical and an inductive
basis for these estimates, how can the estimators defend themselves against
the charge that they're pulling their numbers out of a hat?[6]

There's a line of defense for Drake estimators that isn't discussed in
Mash's otherwise comprehensive analysis. Why not perform an induction
over the educated guesses of successful scientists? The idea is that if the
past guesses of a population of persons or devices have proven to be right
more often than can be accounted for by chance, we have inductive grounds
for accepting subsequent guesses from members of the same population.
It's arguable that success in the empirical sciences requires an intuitive
ability to make good guesses on the basis of unsystematic, holistic considerations—or else the scientist would not be guided to undertake what turns
out to be a fruitful research program. On this account, successful scientists

are conceived to be oracular devices for whose accuracy we have good inductive evidence, even though we lack an account of how they arrive at their estimates. If an eminent astrophysicist guesses that the fraction of stars possessing planets is about 1/10, that's at least *some* evidence that the fraction is about 1/10.

One problem with this induction over successful scientists is that the successful scientists are not of one mind about ETI. There is a pattern, though: the physical scientists (e.g., Sagan) tend to be pro-ETI, while the biologists (e.g., Mayr) tend to be anti-ETI. The indicated procedure is obvious. Successful physical scientists have proven themselves to be good estimators specifically of physical factors, and successful biologists have proven themselves to be good estimators of biological factors. The logic of induction dictates that we should accept the estimates made by each group of the factors relevant to its own area of expertise—the physical scientists' estimates of the physical factors and the biologists' estimates of the biological factors.

Here's why this mode of resolution won't work. Suppose I'm asked to estimate the probability that it will rain on a randomly selected day in Kathmandu. I have no theoretical knowledge of meteorology that bears on this question, nor do I have any detailed frequency information that would enable me to tackle the question inductively. Still, I can come up with a ball-park figure. I remember that Kathmandu is not a desert, so the probability isn't anything as low as .01. In fact, I think it's reputed to have a rainy season of several months during which it rains virtually every day. So the probability of rain on a random day is going to be at least .2. By relying on vague memories and anecdotal reports of this type, I can come up with an estimate that has some rational foundation. At least it would be more rational to accept the estimate than to accept a figure for the probability that's literally drawn out of a hat. It would be exceedingly bizarre, however, if my ruminations based on vague memories and unsystematic anecdotes led me to the estimate that the probability of rain in Kathmandu is .3987. It would be even more bizarre if another person, to whom I had conveyed this estimate, tried to get me to change it to .3986. It's utterly implausible to suppose that the sets of memories and anecdotes that warrant an estimate of .3987 are discriminably different from the sets of memories and anecdotes that warrant an estimate of .3986. To make either estimate is to commit the *error of over-precise estimation*—that is, to specify a quantity with a degree of precision which is unwarranted by the means available for its measurement or estimation.

Now we've seen that the resolution of the ETI question depends on the outcome of a struggle between very large numbers and very small numbers.

Suppose that it's been estimated that there are 10^{22} inhabitable planets in the universe. (This is more or less in line with the estimates of pro-SETI physical scientists). Then, if the frequency of life on an inhabitable planet and of intelligence on a planet containing life are both one in a million, the estimated number of planets with intelligent life would be an enormous 10^{10}, and the rational acceptability of the ETI hypothesis would be assured. But if both those frequencies were taken to be one in a trillion, the estimated number of planets would be zero (to the nearest integer). So the disposition of the ETI hypothesis may very well turn on whether some frequency estimates are one in a million or one in a trillion.

How reasonable is it to suppose that the educated guesses of successful scientists are sufficiently refined to discriminate between a frequency of one in a million and a frequency of one in a trillion? From one perspective, the difference between the two frequencies is very large. The fair value of a lottery ticket which gives us one chance in a million to win is one million times the fair value of a ticket where the chance of winning is one in a trillion. If the prize is a million dollars, the value of the former ticket is one dollar, while the value of the latter ticket is 1/10,000 of one cent. From another perspective, however, the difference between these two frequencies is very small. In fact, the amount by which one frequency exceeds the other is substantially less than the difference between the two estimates of .3986 and .3987 for the probability of rain in Kathmandu. So is the difference between one in a million and one in a trillion very large or very small? The answer depends on one's purpose. If the purpose is to calculate the fair value of lottery tickets, the difference is very large. The purpose here, however, is to determine the ease with which the two frequencies can be *discriminated* on the basis of holistic and unsystematic informational input. For *that* purpose, it seems compelling that the difference between one in a million and one in a trillion is fantastically small. It may be reasonable for us to accept successful scientists' guesses that some particular Drake frequencies are either "large" or "moderate" or "small." Perhaps it's even reasonable to suppose that they can tell us whether a frequency is merely "small" or "very small" or "very, very small." But it's inconceivable that there could be two sets of discriminable holistic tableaus, one of which would warrant the one-in-a-million estimate and the other the one-in-a-trillion estimate. Such a feat of discrimination would be akin to discriminating visually between a million-sided polygon and a trillion-sided polygon. To insist on either value is to commit the error of over-precise estimation. Yet the two values quite likely give different answers to the ETI question. Therefore, we can't expect to obtain a rationally compelling answer to the ETI question based on the educated guesses of successful scientists.

1.3 THE LUCRETIAN ARGUMENT

What about the Lucretian move from mere enormity to literal infinity? While it's true that no finite N is big enough to do the job that's required of it in the one-sided vastness-of-the-universe argument, the situation is changed qualitatively if it's supposed that there are infinitely many occasions for an event. The argument, in a nutshell, is that if the number of occasions for an event is infinite, then no matter how close to zero the probability of its occurrence may be, it's bound to happen. A monkey striking typewriter keys at random forever will eventually type the verbatim text of *Hamlet.* By the same token, if the world is infinite in space or time, isn't it equally inevitable that the random collision of atoms will produce another intelligent species? The modern version of the Lucretian argument comes to an even more radical conclusion than Lucretius. The latter merely claimed that the falsehood of the ETI hypothesis was "in the highest degree unlikely." The contemporary claim is that if the universe is infinite, then there are *bound to be infinitely many worlds with ETIs:*

> Analysis shows that . . . an open universe must be infinite in extent, with an infinite number of galaxies, an infinite number of stars, and an infinite number of planets. In an infinite universe, any event which has a finite probability—no matter how small—of occurring on a single given planet must inevitably occur on some planet. In fact, such an event must occur on an infinite number of planets (Hart, 1982, p. 163).
>
> If cosmological initial conditions are exhaustively random and infinite then anything that can occur with non-vanishing probability will occur somewhere; in fact, it will occur infinitely often (Barrow & Tipler, 1988, p. 7).

These opinions are based on the firm ground of modern probability theory—specifically, the Borel-Cantelli theorem, which specifies that in an infinite number of independent trials, any event with non-zero probability will occur infinitely often (Chow & Teicher, 1978). It's important to note, however, that the Borel-Cantelli theorem doesn't quite underwrite the existence of ETIs in an infinite universe. The firmly grounded thesis is that there are surely ETIs in an infinite universe *if* the probability of ETI is non-zero. But how do we know that the laws of the universe allow for ETIs? The pro-ETI reply is that this antecedent condition is assured by our own existence: the fact that there's intelligent life on earth establishes that the probability of intelligent life developing on a randomly selected planet is greater than zero. This argument is too quick, however. It overlooks the possibility that the laws of nature may specify that certain events occur exactly once—that is, that the probability of some *recurrences* may be exactly zero. This issue came up in the previous section, where a well-behaved universe was described whose laws did in fact entail that certain events were unique. Though it wasn't relevant to mention it at the time, this universe also happened to be temporally infinite. Thus, it isn't necessarily true that every event that occurs once must inevitably repeat itself in an infinite universe.

Of course, the universe described in the previous section was an exceedingly implausible place. Perhaps if we restrict ourselves to universes that satisfy certain plausible conditions we will find that in all *these* places what occurs once must have a non-zero probability of recurring. Mash (1993) lists the following six conditions that must be satisfied if the Lucretian argument is to go through.

1. The first condition is, of course, that the universe must be infinite. It makes no difference whether it's infinite in space or in time. One monkey randomly typing forever will eventually produce *Hamlet,* and infinitely many monkeys randomly typing for a single day will produce *Hamlet* on that day.
2. The number of different kinds of materials in the universe must be finite. "If the keyboard of the Shakespearean typewriter contains an infinite . . . variety of types of keys, recurring sequences need not occur" (Mash, 1993, p. 204).
3. The state of affairs to be repeated must itself be finite. If *Hamlet* were infinitely long, then even a monkey typing forever might never produce it.
4. The laws of nature must be the same everywhere in the universe. Otherwise, the universe may be partitionable into infinitely many finite-sized sub-universes, each of which is governed by different laws. By virtue of its finite size, one of these sub-universes may contain an event which is unique to that sub-universe. By virtue of their operating under different laws of nature, that unique event may be forbidden to occur in the other sub-universes. The result would be an event that occurs only once in the whole universe.
5. All objects must be aggregates of specifiable, discrete elements. It seems to me that this condition of Mash's is tautologously satisfied in every logically possible world. For suppose that some type of object X is *not* describable as an aggregate of other elements. Then X is itself an element. At the very least, every object is an aggregate composed entirely of itself. It's true that the Lucretian doesn't want to add infinitely many types of indecomposable objects to the class of elements, for the existence of infinite varieties means that recurrence isn't assured after all. But this is ruled out by condition 2.
6. Determinism is false. If some processes in the world weren't inherently stochastic, then ETI might be ruled out by deterministic laws even in an infinite universe. "If determinism is thoroughgoing, the typewriter might as easily produce an infinite sequence of 'a's as all possible sequences" (p. 210).

All of Mash's conditions correspond at least to scientifically respectable hypotheses. The most dubious one is the hypothesis that the world is in fact infinite. But even this is by no means definitively ruled out by modern cosmology. So can ETI advocates claim that, given our current scientific understanding, ETI is a sure thing *if* the universe is infinite? The answer is: not yet. Mash shows only that his conditions are *necessary* for the Lucretian argument to go through. He doesn't attempt to show that they're *sufficient.* In fact, they surely aren't sufficient. It's trivially easy to construct universes that satisfy all six Mash conditions, but which contain events that occur only once. The hypothetical universe described in section 2 won't do the job because it's a deterministic world, in violation of condition 6. But consider a world which is infinite in space and time (condition 1), contains nothing other than a finite number of different types of discrete elements (conditions 2 and 5) that move about in accordance with inherently probabilistic laws (condition 6) that are the same everywhere (condition 4). Suppose also that the elements are indestructible and uncreatable: their number is, always has been, and always will be the same. Now there's nothing in Mash's conditions that requires that there be infinitely many particles of each type. Suppose then that there are exactly n particles of type T, where n is a finite number. Define a *Tn-object* as any aggregation of n T-type particles, irrespective of the distances between the individual particles. This object is finite, which satisfies the last remaining Mash condition (condition 3). Yet it seems that there exists, always has existed, and always will exist exactly one T_n-object in the universe. The fact that one of them exists is evidently not a reason for supposing that there's a non-zero probability that another one exists somewhere else. If the reader has qualms about accepting T_n-objects as bona fide objects, then consider a universe just like the one described above, except that it contains only a single T-type particle. Then that particle is eternally unique.

I don't, of course, claim any plausibility for this artificial world, nor do I claim that ETI would be any more problematic in it than in some other imaginable world. The point of the construction is only to show that Mash hasn't yet formulated a set of premises which is sufficient to bear the Lucretian argument to its conclusion. At present, nobody knows exactly what a sufficient set of premises is going to look like. For all we know, the list will include conditions that our universe clearly fails to satisfy. Until we know otherwise, Lucretian considerations don't provide us with rational grounds for believing in ETI, nor even for believing in the conditional proposition that there are ETIs if the universe is infinite.

Moreover, when it comes to the *SETI* hypothesis that we might be able to acquire direct empirical evidence of the existence of ETIs, the infinity of the universe becomes irrelevant. According to relativity theory, we can receive signals from only a circumscribed region of space-time—the region from which signals travelling at the speed of light are close enough to have arrived. If the universe has an infinite past, this "light cone" could theoretically encompass an infinitely large portion of space-time. But here we must reckon with another canonical theory of modern cosmology—the Big Bang theory. The Big Bang theory doesn't logically entail that that the universe has a finite past. It's compatible with the theory, for instance, that the Big Bang was preceded by an era in which the universe contracted to a point of maximal density—a Big Crunch. But if this is so, then the Big Crunch poses an insurmountable informational barrier. It would be impossible for any empirical information about pre-Big-Crunch conditions to survive the crunch. So one way or another, our contemporary scientific views entail that the *observable* universe is a finite chunk of space-time. Therefore, Lucretian arguments, even if they turn out to be sound, provide no support for SETI.

1.4   THE ARGUMENT FROM MEDIOCRITY
AND INDIFFERENCE

There's one more pro-ETI argument to evaluate—the argument based on the *principle of mediocrity*. Despite its frequent employment, no precise statement or adequate analysis of this principle exists. The principle seems to be that we should take it as a default assumption that the object or event under consideration is more likely to be average or typical of its type rather than unusual or unique. In the absence of compelling reasons to suppose otherwise, our best guess is supposed to be that "our surroundings are more or less typical of any other region in the universe," that "there is nothing special about the solar system and the planet earth" (Rood & Trefil, 1981, p. 4), that "our own evolution is typical" (Tipler, 1985, p. 134), and so on. The persistent application of the principle magically transforms our ignorance into a case for the existence of ETIs:

> In estimating n_e, for example, we start by allowing (on the basis of the assumption of mediocrity) that the environmental conditions on many other planets are roughly like those of the Earth four billion years ago. Then we take the next term, f_l, and argue (on the basis of something like Miller and Urey's 1959 "primordial soup" theory) that in such conditions life will likely arise. But now if our theory on this point is incomplete, imprecise, or found wanting in other respects, we may leap the hurdle by simply granting ourselves (again, on the basis of the assumption of mediocrity) that life is not uncommon on planets with Earthlike conditions, and then proceed to f_i, arguing from the basis of natural selection that intelligent life will arise—and so on through the equation. The

assumption of mediocrity thus acts as a kind of fall-back, a temporary (we hope) stopgap that lets us get on with the theory at hand (Mash, 1993, p. 214).

One has to admire the sheer audacity of this grab for credibility. Having failed to arrive at a substantial probability for ETI by appeals to the vastness of the universe, ETI advocates now claim to have achieved their goal *without* relying on the vastness of the universe.

What, exactly, does the principle of mediocrity tell us about probabilities? Here's a first pass at an interpretation: if an item is drawn at random from one of several sets or categories, it's likelier to come from the most numerous category than from any one of the less numerous categories. In the case of ETI, the single case of Planet Earth is drawn from the category of "inhabitable planets containing life," as opposed to the category "inhabitable planet not containing life." Therefore, since the principle of mediocrity tells us that it's likelier to have been drawn from the more numerous category, we conclude that the set of inhabitable planets containing life—the set from which the earth was in fact drawn—is probably more numerous than the set of inhabitable planets not containing life. In point of fact, SETI advocates don't actually draw the inference that the set of life-bearing planets is *more* numerous than the set of non-life-bearing planets. The conclusion is the rather vague claim that the proportion of life-bearing planets is not inconsiderable. Presumably, the principle of mediocrity ascribes progressively smaller probabilities to ever-smaller estimates of the proportion of life-bearing planets, the least likely scenario being that our single case is unique. The criticisms to be wielded below against mediocrity reasoning are applicable, *mutatis mutandis,* to the vaguer and more convoluted principle. But the critique itself would become convoluted, and would make for an unpleasant stretch of reading. So I'll run my discussion on the original, simplified formulation—the one that says that the set from which the single sample is selected is more likely than not to be more numerous than its complement. Everything that's wrong with this claim is also wrong with the vaguer claim.

There are two problems with the ETI argument from mediocrity, both of them utterly debilitating. The first is that whatever *prima facie* plausibility the principle of mediocrity may have is entirely dependent on the single case having been drawn at random. But the earth is not a randomly selected planet:

> . . . we did not "choose" the Earth randomly. It is trivially true that the only planets on which intelligent conscious life like ours can evolve are those on which the conditions for that evolution are met. It is not as though we were like some extraplanetary being walking in on the galaxy from the outside . . . and choosing a planet at random. We have no choice but to start with a planet on which intelligent life has evolved, but that in no way insures that the case is

typical. (Apparently ETI proponents would like to argue that if, e.g., habitable planets were very, very rare, we would have found ourselves inhabiting a "typical" uninhabitable planet rather than the Earth.) So even if the assumption of mediocrity is a valid means for licensing induction from a single case, it may not be applicable for certain crucial inferences along the Drake equation (Mash, 1993, p. 218).

The problem of randomness aside, the principle of mediocrity is demonstrably defective. Recall its formulation: if an item is drawn at random from one of several sets or categories, it's likelier to come from the most numerous category than from any one of the less numerous categories. This statement is amenable to two drastically different readings, one of which is a probabilistic truism, the other a fallacy. The principle that's needed to underwrite ETI is the fallacious version. But the fallacy is obscured by virtue of its being confused with the truism. On one reading, the principle states that the single randomly drawn object is more likely to have come from the category that we know to be more numerous. This is the truism. If category A contains 3 elements and category B contains 1 element, then a random draw from the total population of 4 elements has a 3/4 probability of having come from A, and only a 1/4 probability of having come from B. This inference presupposes that we have antecedent knowledge of the relative numerosities of the classes A and B. In its ETI application, however, our antecedent knowledge and the inference we draw from it are reversed. We know that the random choice has come from A, and we infer from this that A is probably more numerous than B. For example, the classes A and B are "inhabitable planets that contain life" and "inhabitable planets that do not contain life," respectively, and the fact that our single examined case belongs to A is alleged to license the inference that A is probably more numerous than B (more vaguely, that the proportion of A's is not inconsiderable). This is an altogether more speculative inference than the first.

Mash provides an indirect refutation of the mediocrity assumption—that is, he shows that it entails consequences which even the most sanguine SETI advocate would be loath to accept. Suppose that our single case is evidence for the greater numerosity of the class to which it belongs than of its complement. Then why should we only apply it to the categories which are represented in the Drake equation? Once again, our planet contains Velcroclad, skate-boarding teenagers. By parity of reasoning, our planet must be regarded as probably average or typical with regard to the incidence of Velcro-clad, skate-boarding teenagers. Therefore, it's likely that Velcro-clad, skate-boarding teenagers reside on most (more vaguely: many) of the inhabitable planets in the universe. The argument is the same. Indeed, the argument is the same all the way to twin Earths. Since its unconstrained use leads to absurdity, there must be some additional condition for the applicability of the mediocrity principle. But there's no hint of what this additional condition might be in the ETI literature.

Mash's argument is an entirely adequate refutation of the mediocrity principle. But it doesn't provide an analysis of where the reasoning goes wrong. The reasoning goes wrong, I think, by virtue of a tacit appeal to the notorious probabilistic *principle of indifference,* which is known to lead to contradictory conclusions. The conclusion of the mediocrity principle is that A, the category from which our single case is known to come, is *probably* more numerous than its complement B. This probabilistic claim means that, in light of the knowledge that the single examined case comes from A, there are more ways for A to be more numerous than B than there are ways for B to be more numerous than A, relative to some equiprobable set of "ways." The source of the temptation to suppose that this might be so is highlighted in the following minimal scenario. Suppose we know that there are three inhabitable planets in our stellar system, including our own. We know that our own planet contains life, but we have no evidence relating to the existence or non-existence of life on the other two. Given only this information, is it any likelier that there are more life-bearing planets in our system than non-life-bearing planets? If the principle of mediocrity is valid on a galactic scale, it should also be applicable to this local planetary scenario. Our own planet would be at least marginally more typical if at least one of the other planets contained life than if neither of them did. Therefore, if the mediocrity principle is valid, it should be at least marginally more probable that one or both of the other planets contains life than that neither of them does. Is this a valid probabilistic inference?

The argument that we're tempted to make is the following. The two unexamined planets—call them P and Q—may each either harbor life or not harbor life. The four possibilities are therefore: (1) neither P nor Q harbors life, (2) both P and Q harbor life, (3) P harbors life but Q doesn't, and (4) Q harbors life but P doesn't. If (1) obtains, there is only one life-bearing planet (our own), as compared to two non-life-bearing planets. If (2) obtains, there are three life-bearing planets (our own included) and no non-life-bearing planets. If either (3) or (4) obtains, there are two life-bearing planets (ours and one other) and one non-life-bearing planet. In all of these scenarios except the first, the number of life-bearing planets exceeds the number of non-life-bearing planets. Therefore, the probability that the set of life-bearing planets is greater than the set of non-life-bearing planets is 3/4.

This conclusion obviously depends on the assumption that the four scenarios are equiprobable. Why should we make that assumption? Well, we've assumed that we have no knowledge relevant to the question whether the unexamined planets contain life. So the judgment that the four scenarios are equiprobable can't be based on any relevant evidence. It must, in fact, be based on our very lack of knowledge. There is no other candidate. We must be assuming that in the absence of any reason for preferring one hypothesis over another, we should regard them as equiprobable. This is the principle of indifference.

Unfortunately for ETI and SETI, the principle of indifference is invalid. The problems of indifference reasoning are discussed extensively by Salmon (1966). The root problem is that being in a state of epistemic indifference between complementary hypotheses A and B is entirely compatible with also being indifferent with regard to hypotheses A, B_1 and B_2, where B_1 and B_2 are an exhaustive decomposition of the state of affairs indicated by B. The indifference principle thus stipulates that the probability of A is 1/2 by virtue of the toss-up between A and B, and that it's 1/3 by virtue of the toss-up between A, B_1, and B_2. More generally, the existence of a state of indifference between A and B is compatible with the existence of states of indifference between A, B_1, B_2, . . . , and B_n, for any n. For example, suppose that I think I have no basis whatever for deciding whether any gods exist, or if any do exist, what their number might be. To avoid having to deal with inessential complications, suppose that this uncertainty over the number of gods extends from 0 to 9: I'm sure that the number of gods is in the single digits, but I have no idea what that digit might be. This is surely a logically possible epistemic state. What does the indifference principle tell us about this state? I'm indifferent between the hypothesis that there are no gods and the hypothesis that there are some gods. Therefore the probability that there are no gods is 1/2. But I'm also indifferent between the ten hypotheses of the form "There exist(s) exactly n god(s)," where n runs from 0 to 9. Therefore the probability that there are no gods is 1/10. Contradiction

It's worth noting that the conjunction of the two states of indifference that lead to conflicting probabilities is sometimes more than merely possible: *being in one of the states sometimes logically entails being in the other.* This is the case in the foregoing example. For suppose I encounter a bit of evidence which asymmetrically supports the hypothesis that there are no gods over the hypothesis that there are some gods. Then it follows that I'm also no longer indifferent between the ten hypotheses of the form "There exist(s) exactly n god(s)"—for my single bit of evidence favors the hypothesis with n = 0 over each of the other nine hypotheses. Thus, indifference over the ten numbered hypotheses *entails* indifference between existence and non-existence. The contradictory conclusions to be drawn by the indifference principle are not only available—they're required. The fact that some states of indifference entail other indifferences that lead to contradictory probabilities is well-known in the literature of probability theory, where it goes by the name of the

Bertrand paradox (von Mises, 1957, p. 77). In retrospect, it's not surprising that indifference reasoning would turn out to be untenable—it's just too good to be true to suppose that complete ignorance can be transformed into precise probabilistic information:

> Knowledge of probabilities is concrete knowledge about occurrences; otherwise, it is useless for prediction and action. According to the principle of indifference, this kind of knowledge can result immediately from our ignorance of reasons to regard one occurrence as more probable than another. This is epistemological magic (Salmon, 1966, p. 66).

Let's return to the scenario of the three planets. It's assumed that we have complete uncertainty regarding whether planets P and Q contain life. It follows that we're indifferent as to the four indicated states of affairs: (1) P and Q contain life, (2) neither P nor Q contains life, (3) P contains life but Q doesn't, and (4) Q contains life but P doesn't. Applying the principle of indifference, we obtain a probability of 3/4 for the hypothesis that there are more life-bearing than non-life-bearing planets. But our complete uncertainty also entails that we're indifferent as to the hypotheses (1) that there is life on other planets and (2) that there is no life on other planets. Applying the principle of indifference to *this* decomposition of the possibilities, we obtain a probability of 1/2 for the hypothesis that there is life on other planets. But the 1/2-probability hypothesis that there is life on other planets is identical to the previous 3/4-probability hypothesis that there are more life-bearing than non-life-bearing planets, which is impossible. The problem with the principle of indifference isn't merely that it gives an *erroneous* probability estimate—it's that it doesn't given any determinate estimate at all.

Moreover, if we're going to permit ourselves to use the principle of indifference, then mediocrity arguments—considerations of our "typicalness"—are supernumerary. We could simply plead complete ignorance of any factor that might bear on the truth or falsehood of the ETI hypothesis, including whether or not our planet, our evolution, etc., are typical or unique. Since there's no reason to favor either the ETI hypothesis or its negation, we could claim that the principle of indifference gives us a probability of 1/2 for each one—certainly high enough to warrant a major expenditure of resources! If the *prima facie* plausibility of the mediocrity principle rests, as I claim, on a tacit appeal to the principle of indifference, then ETI advocates might as well skip the mediocrity trimmings and just make a direct appeal to indifference. If the argument from mediocrity were valid, it would be superfluous.

To summarize: there are no arguments on the table that would underwrite any degree of optimism—or pessimism—about the ETI hypothesis. The same applies to the stronger SETI and CETI hypotheses. The conclusion isn't that we should repudiate these hypotheses. It's that we have no grounds for having any opinion at all. A few words of clarification on what this epistemic position is like. So long as we describe the epistemic options in terms of the dichotomous folk-psychological concept of belief, the distinction between disbelief and agnosticism is clear enough. To disbelieve a hypothesis X is to believe that its negation not-X is true; to be agnostic about X is neither to believe X nor to believe not-X. When we talk in terms of probabilities, however, the agnostic option of suspending judgment may seem to be lost. What is the probability of X for an agnostic? To say that it's 1/2 is to appeal illegitimately to the principle of indifference. But there's also no warrant for making the probability of X either greater than or less than 1/2. To make sense of agnosticism, we have to make a conceptual space for *vague* probabilities.[7] Hypotheses can't always be associated with discrete probabilities. Sometimes the best we can do is to ascribe a probability *interval* to the state of affairs under consideration. For example, we may know nothing more than that X is likelier to be true than false. This epistemic state of affairs would be represented by ascribing the vague probability (1/2, 1] to X. (A parenthesis indicates that the end-point is excluded from the interval, while a bracket indicates that the end-point is included in the interval.) To say that we have no idea whether X is true or false is to associate X with the entire probability interval [0, 1]. My assessment of the state of the three extraterrestrial hypotheses can be represented by saying that there's no basis for associating anything less with them than the full probability interval [0, 1].

Chapter 2

On the Pursuitworthiness of Extraterrestrial Studies

2.1 DECISIONS UNDER CONDITIONS OF PERVASIVE UNCERTAINTY

In this chapter I'll take a closer look at the metaquestion whether ETI is a project worthy of intensive pursuit. In the previous chapter I tried to ascertain the probability that the pursuit will meet with success. But obviously, the answer to the metaquestion depends on both the likelihood of success and on how interesting or important a successful outcome would be. A high probability of success wouldn't make a project pursuitworthy if the cost were enormously high and the payoff relatively puny. For example, we'd have a high probability of succeeding if we were determined to record the license plates of all the cars that pass by a randomly chosen crossroad in Little Rock, Arkansas, for a period of twenty years. But that doesn't mean it's a good idea to do it. Conversely, a low probability of success—even a very low probability of success—may be compatible with pursuitworthiness if the potential payoff is enormous.

How do probabilities and values combine to determine a rational decision? This isn't the place to entertain novel proposals for measures of program evaluation. Throughout this book, my aim is to ascertain where ETI stands in relation to the currently received canons of scientific rationality. I will approach the metaquestion from the standpoint of traditional decision theory. Or rather, I'll do so as much as I can. It will turn out, however, that the pervasive absence of information with which the decision has to be made forces us to introduce some conceptual extensions to the standard decision—theoretical framework erected by von Neumann and Morgenstern. However, these extensions are natural, inevitable, and I think uncontroversial.

In decision theory, it's conceived that the decision-maker is confronted with a number of mutually exclusive courses of *action.* The task is to provide rational grounds for selecting one of them. Each possible action is conceived to have any number of possible *outcomes,* and each outcome has assigned to it a *probability* and a *value.* Let $x_1, x_2, \ldots, x_n$ be the n possible outcomes of action X. Then, $p(x_i)$ is the probability that x_i will occur as a consequence of choosing to do X. The value $v(x_i)$ of x_i is a measure of the relative preference for this outcome as compared to others. There are well-known decision–theoretical techniques for assigning numbers to values up to an *interval* scale, which is to say that the zero-point and unit value are arbitrarily selected, whereupon all the other values are fixed. Decision theory stipulates that the best course of action is the one with the highest *expected value* (EV), this quantity being the sum of the products of the probabilities and values of all that action's outcomes:

$$EV(X) = p(x_1)v(x_1) + p(x_2)v(x_2) + \ldots + p(x_n)v(x_n).$$

Why should we maximize EV? Because a persistent policy of doing so results in a greater realization of value than any other policy in the long run.

In most cases of real-life decision-making, one of the available courses of action is *inaction*—doing nothing at all. Inaction can be treated as just another action to which an EV is assigned on the basis of its possible outcomes and their values and probabilities. The outcomes of inaction may be substantial. If a gunman threatens to shoot you unless you play either game X or game Y, the EV of inaction will undoubtedly be less than EV(X) and EV(Y), and you would do well to choose a game and play it even if they're both boring. If no particular consequence is expected to follow from inaction, then inaction is a choice with a single outcome—the continuation of the status quo. The continuation of the status quo will of course have a value of its own, which may be either higher or lower than the values of the outcomes associated with other available courses of action. For example, the status quo may include the fact that I have a toothache, in which case inaction could have a lower EV than the option of going to the dentist.

Choices with a single outcome come up in other ways too. For instance, you may be given a choice between playing a game of chance and receiving a certain prize outright. If you opt for the outright prize, then it's yours for sure. In such a case, the action of accepting the prize has a single outcome x, whose probability $p(x)$ of occurring if that action is taken is 1. Thus the expected value of that choice is:

$$EV = p(x)v(x) = 1v(x) = v(x).$$

That is to say, the expected value of a single-outcome choice is identical to the value of that outcome. If inaction has no other possible outcome except the continuation of the status quo, then the expected value of inaction is the value of the status quo. Since the zero point of value is arbitrary, it's often convenient to assign the status quo a value of zero.

The analysis of the ETI hypothesis in the previous chapter led to the conclusion that its probability is radically vague. How do vague probabilities fit into the decision-theoretical picture? Consider a game of chance X where the probabilities of the two possible outcomes—call them w and l for "win" and "lose"—are totally vague. This means that $p(w)$ may be any number between 0 and 1. Whatever $p(w)$ may be, it has to be the case that $p(w) + p(l) = 1$. So while both $p(w)$ and $p(l)$ can range from 0 to 1, they can't both be very low or very high simultaneously. Now suppose that the outcome w has a greater value than the outcome l (that's why we call it "w"). Then the indeterminacy in probabilities entails a corresponding indeterminacy in the expected value of the game X. The highest value that $EV(X)$ can have is obtained when $p(w) = 1$ and $p(l) = 0$. In this case,

$$\max EV(X) = p(w)v(w) + p(l)v(l) = 1v(w) + 0v(l) = v(w).$$

The lowest value that $EV(X)$ can have is obtained when $p(w) = 0$ and $p(l) = 1$, whereupon

$$\min EV(X) = p(w)v(w) + p(l)v(l) = 0v(w) + 1v(l) = v(l).$$

Thus $EV(X)$ can be anything between the value of winning and the value of losing. If we can't be more precise about probabilities, then we also can't be more precise about expected values.

Nevertheless, this information may be enough to make a rational decision. If you play game X, the worst-case scenario is that its EV is equal to $v(l)$, the value of losing. If your other options all have EV's *less* than $v(l)$, then X is the indicated choice, despite the radical indeterminacy in its expected value. For example, if somebody offers to play a (painless) game with you in which you receive \$2 if you win and you receive \$1 if you lose, you would do well to accept the offer even if you know nothing at all about the odds of winning. This configuration of circumstances actually comes up in the evaluation of the pursuitworthiness of ETI. We'll see that some ETI advocates maintain that the pursuit of ETI will pay off handsomely even if it fails—for failure, like success, would tell us some very important things about ourselves and the universe we live in. Call this the *can't-lose* hypothesis. If the can't-lose hypothesis is true, then of course we can answer the metaquestion in the affirmative even though the probabilities of ETI success or failure are entirely indeterminate.

It's also possible to make a decision in the face of probabilistic indeterminacy if the values of both the winning and the losing outcome are less than the expected value of another available alternative. In this case, the decision is that we should *not* undertake the indicated course, for we can do better no matter what the outcome turns out to be. In all other cases, however, probabilistic indeterminacy robs decision theory of the basis for making a recommendation. Suppose that the value of winning is higher than the expected value of the best alternative action, but that the value of losing is lower than the expected value of the worst available alternative. Then there's no way of knowing whether we should play the game. The decision has to be made non-rationally.

We're almost ready to assess the values of success and failure at ETI. There's just one more preliminary issue: an argument by Regis (1985) to the effect that failure is not one of the possible outcomes of engaging in ETI research. Regis maintains that we can never arrive at an epistemic state from which it's appropriate to say that ETI has failed. His point isn't merely that the probability of failure, like the probability of success, is radically indeterminate. It's that the failure of ETI can never be established, no matter what additional information may come our way in the future. In this respect, there's an asymmetry between ETI success and ETI failure. However likely or unlikely they may be, there are scenarios that would persuade us of the truth of the ETI hypothesis. Suppose, for example, that we track a spaceship travelling from beyond the orbit of Pluto until it lands on earth and that five-legged beings come out of the ship and begin to converse with us in English, explaining that they come from a planet circling the star of Sirius and that they've learned English by analyzing our old radio and TV broadcasts. Now there may be good reasons for believing that this scenario will never be actualized. But if it *were* to be actualized, then we would know that ETI is true. Regis claims that we can't in a similar manner envision an ascertainable state of affairs that would definitively indicate that ETI is false:

> The problem is that even a major search's failure to detect an interstellar message would not by itself disprove the existence of extraterrestrials. For no matter how extensive and complete a search may be, the possibility would always remain that it was not complete enough, that maybe we had not searched the right places at the right times, or on the right frequencies, or with the right reception media. There are countless things that we might have missed, or even have failed to look for: gravity waves, tachyon messages, neutrino modulations, and so on and so forth. Since there is no way of knowing that any given search—or series of them—has exhausted all the possibilities, there is always the residual chance that undetected aliens are still out there, that whole civilizations have escaped our notice . . . [I]f we make a search and end up only with absence of evidence for ETI, then the only thing we would be entitled to conclude from this is that we have no evidence of their absence. In other words, we are no better off—and no

worse off either—than we were when we started: we still don't know if they're out there (Regis, 1985, p.233).

If Regis is right, then the two outcomes that jointly determine the expected value of ETI aren't success and failure, but success and no resolution at all. Contrary to the suggestions of some ETI advocates, non-success wouldn't teach us anything about the rarity or preciousness of intelligent life. It wouldn't teach us anything at all. The alternative to success is in fact a continuation of the epistemic status quo. But the continuation of the status quo is free. Therefore, unless there is some prospect for success, ETI isn't worth a dime.

The confirmational problem that Regis points to affects all non-existence claims in the same way. A hypothesis to the effect that a certain type of entity or process *exists* might come to be firmly established on the basis of a single serendipitous observation. This isn't the case with the hypothesis that a certain type of object *doesn't* exist—for no matter how many places and times the object is observed to be absent, it remains possible that it exists somewhere else or at some other time. Nevertheless, Regis's conclusion is too severe. Despite the difficulty of verifying negative existential claims, many propositions of this type make their way into science. Contemporary science affirms the non-existence of the luminiferous ether, phlogiston, unicorns, Abominable Snowmen, perpetual motion machines, etc. Some of these beliefs are based on ordinary inductive reasoning. While it's true that we haven't examined every place where Abominable Snowmen might lurk, the fact that they haven't been located in so many of the places where they might have been gives us inductive grounds for supposing that they aren't in any of the unexamined locales either. To be sure, we might be wrong: inductive inference always brings the possibility of error in its train. But inferences to universal *absences* are not riskier than inferences to universal *presences*. The move from the absence of Abominable Snowmen or ETIs in the places we've looked to their absence in places we haven't looked is on the same epistemic footing as the move from the blackness of observed ravens to the blackness of unobserved ravens.

More importantly, Regis doesn't take into account the possibility of a *theoretical* resolution to the ETI question. Let's grant that our failure to find ETIs can never warrant the belief that they don't exist. Even so, we might arrive at the conclusion that ETIs don't exist on the basis of a theoretical argument. Theoretical reasoning is the basis for many of the non-existence claims made by scientists. Physicists' conviction that there can never be a perpetual motion machine isn't primarily based on an inductive generalization from the fact that nobody has ever succeeded in building one—it's based on the laws of

thermodynamics. By the same token we may come to have theoretical reasons
for supposing that ETIs don't exist either, or that we can never find them even
if they do exist, or that we can never communicate with them even if we find
them. Whether by brute induction or by subtle theoretical reasoning, it's conceivable that we might rationally come to the conclusion that one or the other
of the extraterrestrial projects is an irremediable failure.

2.2   THE METAQUESTION

What are the sources of value in ETI? I'll deal with this question in two
stages. First I'll erect a rough-and-ready taxonomy of relevant types of value;
then I'll assess how much of each type is actually possessed by ETI.

To begin with, there's the *epistemic* value of ETI—the value of what ETI
success or failure can teach us about the world. The main contrast to epistemic value is the *pragmatic* value of gifts other than knowledge that may be
bestowed on us by well-intentioned aliens: new technologies, new art forms,
new jokes, etc. The gift of new sciences is intermediate between the two types
of value. It's like epistemic value in that it amounts to the acquisition of new
knowledge. But it's unlike other sources of epistemic value in two respects,
one of which is important to the assessment of the metaquestion. By definition, the prototypical epistemic value is the value of the knowledge that *we*
derive by exercising our rational faculties on the datum that ETI (or SETI, or
CETI) succeeds or fails. Receiving the gift of a new science is obviously not
the same thing. But this is the unimportant difference. The reason that the gift
of new science should be distinguished from prototypical epistemic values is
that the former is not going to be forthcoming unless we succeed at *CETI* as
well as SETI and ETI: we can't be given new sciences unless we're able to
communicate with the would-be donors. In this regard, the gift of new sciences is akin to the gift of new art forms and jokes. In contrast, prototypical
epistemic values may already be available with mere SETI success, or even
with bare-boned ETI success: the bare fact that ETIs exist may already have
major implications for our view of the world. For the purpose of assessing the
metaquestion, the gift of new sciences will be grouped together with the gift
of new jokes as a source of pragmatic value.

Another important distinction is drawn by Regis (1985). There's a difference between the epistemic consequences that can rationally be drawn from
ETI success or failure and the cognitive consequences that will in fact ensue,
regardless of whether they're rationally warranted. For example, it's sometimes been suggested that ETI failure would produce the positive benefit of
causing us to regard intelligent life as more precious than we currently do

(Sagan, 1983). It's not at all clear that this is a rational inference. It seems to be based on the principle that the value of an intelligent life is a function of the total number of available intelligences. But it's a consequence of *this* principle that the murder of an individual human being becomes less reprehensible as the total population of the world increases. If we're not prepared to accept this consequence, then why should we view the non-existence of ETIs as relevant to the value of human life? This issue of moral theory aside, however, we might very well all agree that anything which increases the preciousness of human life is to be counted as advantageous. So if the non-existence of ETIs has that effect, it's to be counted as a benefit that accrues to ETI failure, regardless of whether the effect is rationally justifiable. The distinction between this type of value and pragmatic value is that the latter involve the acquisition of *capacities* that we may or may not choose to exploit. The fact that we're given new sciences or new technologies or new jokes doesn't automatically entail that we have to accept, utilize, or tell them, respectively. To the extent that the gift is only pragmatic, its employment or non-employment is up to us. In the case of ETI failure leading to an increase in the estimated preciousness of intelligent life, however, the former stand in relation to the latter as cause to effect. Beneficial or harmful causal effects of ETI success or failure will be called *causal* values. In the example of ETI failure leading to an increase in the perceived preciousness of human life, the causal effect is cognitive: ETI failure produces a beneficial change in our mentality. The causal effects of the various ETI and SETI outcomes may also be non-cognitive. For example, SETI success may lead the contacted extraterrestrials to fly to earth and enslave us. If this is a real possibility, it would have to be considered in our overall assessment of the advisability of funding SETI projects. In my terminology, it would be a non-cognitive source of causal value.

Finally, there's the intrinsic interest in gratifying our curiosity about ETI, independently of its bearing on any broader issues or of its causal or pragmatic consequences. As Regis says:

> SETI searches do not have to be justified on the grounds of legendary benefit . . . in order for it to be worth our while to continue searches . . . For there will be one undeniable benefit if one day we have a positive result: it will satisfy our curiosity to know what else is out there (p. 243).

I call this the *gossip* value of ETI success or failure. Gossip value is a special case of epistemic value. It's the minimal amount of epistemic value that the discovery of any new fact always possesses. Some new facts possess enormous epistemic value by virtue of leading us to have new insights into the

workings of the universe. They may, for example crucially confirm or disconfirm a fundamental theory. Other new facts may not have much of a bearing
on our other opinions. But every new fact possesses at least the epistemic
value of causing us to learn that it itself is true. This irreducible minimum
of epistemic value is that fact's gossip value. Different facts have different
amounts of gossip value by virtue of being intrinsically more or less interesting. Everyone will undoubtedly agree that the gossip value of the putative fact
that there are (or aren't) other intelligent beings in the universe is greater than
the gossip value of the fact that the per capita consumption of cheese in the
United States was (or wasn't) 24 pounds in 1987.

Now let's consider cases: what are the values of success and failure at ETI,
SETI, and CETI? I've divided the topic into three parts: (1) the pragmatic and
causal values of failure, (2) the pragmatic and causal values of success, and (3)
the epistemic and gossip values of both failure and success. First, what are the
pragmatic and causal values of ETI failure? Since pragmatic values attach to
capacities, there's no such thing as pragmatic harm. We might be indifferent to
the acquisition of a capacity, but it can never diminish our fortune. If extraterrestrial humor causes us to cringe, we simply needn't indulge in it. So when we
canvas the sources of value in the several extraterrestrial projects, we need only
consider pragmatic values on the benefit side of the ledger. But it's clear that
failure at any of the extraterrestrial projects isn't going to result in any increased
capacity. Thus pragmatic values are irrelevant to ETI, SETI, or CETI failure.

What about causal benefits? Some candidates have been floated. According to Sagan, ETI failure "would tend to calibrate something of the rarity
and preciousness of the human species" (1983, p. 36); it "would underscore,
as nothing else in human history has, the individual worth of every human
being" (1980, p. 314); it would have a "sobering influence on the quarrelsome nation-states" (1982, p. 33). You get the idea. Presumably, this benefit
would require full ETI failure. The more moderate failures of SETI or CETI
wouldn't have the same sobering effect, since they wouldn't show that we're
alone in the universe.

Regis has two counterarguments against Sagan's claim, one of which
strikes me as persuasive. Both arguments make use of the historical fact that
in pre-Copernican times, European society was firmly convinced that human
beings were the sole intelligent species in the universe. Regis notes that Sagan
condemns this pre-Copernican view as a "deadly chauvinism" (Sagan, 1983,
p. 30). But how can the belief in our uniqueness be a good thing now if it was
a bad thing then?

[I]f it was wrong for the pre-Copernicans to have such distastefully ethnocentric,
chauvinistic, and pridefully self-important notions about themselves . . . then how

is it somehow beneficial to return to those same notions once the possibility of extraterrestrials is ruled out? This would seem to be having our cake and eating it too (Regis, 1985, p. 234).

Regis has a point. But there are charitable ways to read Sagan on this issue. Sagan could consistently maintain that the knowledge that we're alone in the universe would have a salutary effect, but that it's a deadly *epistemic* chauvinism to be certain that we *are* alone in the universe. On this account, Sagan condemns the pre-Copernicans for accepting an improbable thesis, but he avers that if the improbable thesis were to turn out to be true, it would have good effects.

This reading of Sagan's thesis saves it from Regis's first charge of inconsistency. But it lays the thesis wide open to Regis's second charge of being counter-indicated by the facts in our possession. For the pre-Copernicans provide us with evidence of what the causal consequences actually are of believing that we are unique in the universe. And the evidence can hardly be said to support Sagan's conjecture:

> [I]f we examine the claim that failure to find aliens "would have a sobering influence on the quarrelsome nation-states," we would have to wonder why the pre-Copernican conception of man's specialness failed to have this effect in years gone by. If, as seems to be true, the nations of those times were not more law-abiding and peaceful than modern states, then what reason do we have for imagining that a return to the old belief in humanity's specialness would have any such sobering influences today? This seems to be just wishful thinking (Regis, 1985, p. 234).

A Saganite might protest that conditions are so different nowadays that there's little reason to believe that our reactions to the acceptance of human specialness would be the same as the pre-Copernicans.' That may very well be. The evidence is far from conclusive. Nevertheless, it's true that what evidence we have, imperfect as it is, speaks against Sagan's hypothesis. But then a rational person should guess that Sagan's hypothesis is false. At best, it's a conjecture that strikes us as not entirely implausible. Now intuitive non-implausibility is *something*—all other things being equal, we should probably prefer a hypothesis that isn't intuitively implausible to one that *is* intuitively implausible. But in this case, there are equally plausible hypotheses that lead to contrary conclusions. For example, we might not implausibly guess that the discovery that we're alone in the universe would precipitate humanity into a state of neurotic depression, loneliness, and anomie. The historical evidence of pre-Copernican behavior doesn't support such a hypothesis, but that still leaves it in no worse an epistemic state than Sagan's more optimistic view.

But what is happening to our canvass of values? Decision theory tells us that the rational advisability of a course of action depends on both the probabilities and the values of the outcomes. We had dealt with probabilities in chapter 1, and now we were supposed to turn to considerations of value. But we find ourselves talking about probabilities once again—for example, the likelihood that humanity's reaction to ETI failure would be an increase in the preciousness of human life or a suicidal depression. What's happening is that the idealizations of classical decision theory are proving to be inadequate for the representation of real-life decisions. Decision theory presumes that the values of the several possible outcomes are fully determinate: if you succeed, you definitely get $5, and if you fail, you definitely lose $1. But here the values of the outcomes—ETI success or ETI failure—are themselves uncertain, on account of our uncertainty concerning the *consequences* of success or failure. Since there's some question whether ETI failure will result in an increase in the preciousness of human life or a suicidal depression, we can't simply assign a value to this outcome. To deal with this more complex situation, we have to extend the conceptual apparatus of decision theory. We have to regard the outcomes themselves as having *expected* values, the calculation of which depends on the probabilities and values of their several possible consequences.

So what's the expected value of ETI failure? Let's look at a single possible consequence of failure: that there will be an increase in the preciousness of human life. Call this consequence C_1. The contribution of C_1 to the value of ETI failure is equal to the sum of two factors: (1) the probability $p(C_1)$ that C_1 will occur if ETI fails multiplied by the value $v(C_1)$ of that consequence, and (2) the probability that C_1 will *not* occur times the value of its *non*-occurrence. A similar pair of factors determines the contribution of the consequence C_2 that there will be a planet-wide depression. We may suppose that there are other states of affairs C_3, C_4, . . . , C_n, which are also possible consequences of ETI failure. In all these cases, it's appropriate to consider the non-occurrence of the consequence as equivalent to the continuation of the status quo and to assign it a value of zero. If we do this, and if we assume that the several outcomes are probabilistically independent of one another, then the expected value of ETI failure can be represented as the sum of the products of the probabilities and values of each possible consequence:

$$EV = p(C_1)v(C_1) + p(C_2)v(C_2) + \ldots + p(C_n)v(C_n).$$

Of course, to assign discrete values to consequences like "an increase in preciousness" and "depression" is still to idealize the real situation—for the values of both these consequences obviously depend on the *degree* of preciousness or depression that's at issue. It's one thing if ETI failure produces universal and permanent disarmament; it's quite another thing if it merely affects the sentiments expressed in print by scientists and literati. If it were worth our while, this situation could be rigorously represented in the decision-theoretical scheme by drawing continuous functions that map degrees of preciousness to probabilities in such a way that the integral of the function over all degrees of preciousness is equal to 1. But given the rough-and-ready nature of our assessment of the relevant probabilities, to follow such a procedure would be to swat a conceptual fly with a conceptual sledge-hammer. For our purpose, the several consequences of ETI failure can be represented as dichotomous: either the preciousness of human life increases or it doesn't, etc. Furthermore, we needn't include *all* the possible consequences of ETI failure in our considerations. It's enough if we take into account those that have a non-negligible probability of occurring *and* that have a substantial positive or negative value (taking zero as the value of the status quo). For example, it might become clear that ETI failure will quite likely produce an increase in the frequency with which words starting with the letter 'f' are used by speakers of English. But who cares?

So what are the probabilities and values of C_1, C_2, and any other relevant consequences of ETI failure? As for their probabilities, we don't face the same sort of blanket indeterminacy as we did when we tried to assess the likelihood of ETI failure itself. For what we're trying to estimate the probability of isn't a hypothesis about utterly unknown extraterrestrial conditions—it's a hypothesis about *our own* reactions to certain beliefs, which have in fact previously been held by some societies in some eras. I refer once again to the pre-Copernican cosmology of Europe. The evidence is that this society's belief in the uniqueness of human intelligence didn't produce an increase in the perceived preciousness of human life, or in Weltschmerz, or in any other social good or evil. Once again, the evidence is far from conclusive. But it's the only basis we have for formulating an opinion about the $p(C_i)$'s. The best hypothesis is thus that none of the C_i will happen. But if they won't happen, then we don't have to agonize over what values we should assign to them: the expected value of ETI failure is going to be zero regardless of what the values of the C_i are.

What about SETI and CETI failure? Well, ETI failure *entails* SETI failure—we won't succeed in contacting extraterrestrials if none of them exists. Thus any source of value that accrues to SETI failure must also accrue to ETI failure. But we've just seen that *no* pragmatic or causal values accrue to ETI failure. Therefore, no pragmatic or causal values accrue to SETI failure either. Similarly, SETI failure entails CETI failure. A reiteration of the same argument shows that CETI failure has no pragmatic or causal value. The grand conclusion is that (epistemic and gossip value aside) Sagan's "can't-lose" hypothesis must be rejected. The only hope for establishing the worthiness of any of the extraterrestrial projects is by an appeal to the benefits of their success. If we expend any resources on the extraterrestrial projects and fail, we will have lost our investment.

Let's turn now to the value of success. I'll start with the minimal case of ETI success without accompanying SETI or CETI success—that is, just finding out that there are ETIs, without actually coming into contact with them. In this minimal case, there are no pragmatic benefits to be obtained. However, Sagan argues that ETI success would have the causal benefit of providing "a powerful integrating influence on the nations of the Planet Earth. [It] would make the differences that divide us down here on earth increasingly more trivial" (Sagan, 1982, p. 32). This putative benefit seems to be pretty much the same as the putative benefit of ETI *failure,* which Sagan surmised would have a "sobering influence on the quarrelsome nation-states" (1982, p.33). Regis argues that Sagan can't have it both ways: if the discovery that we're alone in the universe diminishes the likelihood of war, then shouldn't the discovery that we're *not* alone *increase* the likelihood of war?

> After all, if we should blow ourselves to pieces here on earth, there is another whole civilization out there waiting to take up where we left off. Made complacent by the knowledge that other intelligences exist, we can play fast and loose with the fate of the earth (Regis, 1985, p. 237).

Sagan's conjectures and Regis's critique can be understood either epistemically or causally: it may be claimed that ETI success (or failure) would provide pacifists with additional support for their position, or that it would *cause* us willy-nilly to become more pacifistic. Regis's objection is telling against the epistemic thesis, although Regis doesn't provide us with all the steps of the argument. He seems to suggest that Sagan's position is logically inconsistent—that you can't, on logical grounds alone, assert that Q is a consequence of both P and not-P. In fact, you *can* assert that Q is a consequence of both P and not-P. But if you do, then the truth-value of P is irrelevant to your acceptance of Q—you should accept Q even if you're totally in the dark about P. If both the discovery that we're alone and the discovery that we're not alone rationally entail pacifism, then we should be pacifists right now, since one or the other of these two alternatives is bound to be correct. So far as the pacifism issue is concerned, there's nothing to be gained by waiting to find out whether we're alone. ETI makes no difference to the case for pacifism.

What if we interpret Sagan as making the causal claim that both ETI success and ETI failure would incline us, whether rationally or irrationally, toward pacifism? In this case, Regis's objection is inconclusive. Pointing out that Sagan assigns the same effect to ETI success as he does to ETI failure may serve to undermine the intuitive plausibility of Sagan's pair of conjectures. But it's at least logically possible that both causal conjectures are true—that both the discovery that ETIs exist and the discovery they don't exist might have a pacifying influence on the quarrelsome nation-states. But the causal conjecture about ETI success is liable to the same pair of objections as the causal conjecture about ETI failure. First, it's no more plausible than the opposite guess that the discovery of ETIs will, as Regis suggests, make us more complacent about blowing ourselves to pieces. Second, what evidence we have doesn't support either guess:

> We have prior experience to guide us here as well, the finding of new worlds right here on earth. The discovery of the Americas, for example, did not have anything like the effect upon Europeans that SETI advocates insist that discovery of ETI will have upon us. It did not make differences among Europeans more trivial, it did not serve as an integrating influence among them, it did not make them more tolerant and peace-loving (p. 237).

So, epistemic and gossip value aside, our best guess has to be that ETI success without accompanying SETI or CETI success has no value. Since ETI failure is also valueless, the conclusion is that the minimal project of ETI sans SETI or CETI is worthless.

On the face of it, it's CETI success that might secure for us a pragmatic bonanza of new sciences and new arts. However, Regis argues that the pragmatic value of CETI success is vanishingly small. For suppose that we receive a message which conveys an account of a new ethical or political system. How shall we assess its worth?

> The only way to assess alien moral or political systems is to compare them to the systems of earth, and to project the consequences of adopting their doctrines here. Such an analysis would, in the end, lead us to embrace those aspects of their ethics that we thought would further human well-being, and to reject those that would not. But in doing this we would be using our own pre-contact standards. Call it provincial, reactionary, chauvinistic, or what you will, we'd be evaluating other theories by reference to our own. But to do this, to use our own accepted moral or political standards as criteria for the evaluation of other, alien ideologies, is hardly to get outside the earthbound perspective from which alien communications are supposed to liberate us (p. 240).

And so it goes, presumably, with other potential pragmatic benefits: we'll never learn anything radically new from ETIs because we'll have to evaluate everything they tell us by our own terrestrial lights. The tip-off that there's something wrong with this argument is that its import is far too strong.

Substitute two arbitrary parties for earthlings and aliens, and it becomes an argument for the proposition that nobody can ever learn anything new from anybody else. Here's a quick (but far from complete) account of what's wrong with Regis's argument. New cultural forms, whether moral, political, aesthetic, or scientific, arise in two stages: first somebody comes up with a new idea, and then the new idea is evaluated as worthy of implementation. All of Regis's remarks address the second evaluative stage. Let's suppose that he's right in claiming that our evaluative criteria are unalterable. It doesn't follow from this that extraterrestrials can't teach us anything new, for they may still present us with new ideas that have both of the following properties. First, by virtue of their different biology, different planetary environment, and different history, certain ideas might occur to ETIs that would never spontaneously occur to any of us humans. Second, when these ideas, which we ourselves would never adduce, are presented to us by an external agency, we might find that some of them rank very high *according to our own system of evaluation.* Nevertheless, we might never have thought of them ourselves. Assuming that CETI succeeds, there's no reason to believe that we couldn't profit from a Plutonian's distinctive take on the universe in this way. Of course, we can't be sure that ETIs will be both willing and able to provide us with pragmatic benefits. But since there can be no pragmatic harm (pragmatic values being capacities), we have everything to gain and nothing to lose by CETI success—pragmatically speaking.

On the other hand, CETI success could have catastrophic *causal* consequences that counterbalance the possible pragmatic benefits. If the ETIs make contact with us, they might come to earth in order to enslave us or cook us for dinner. Considering both pragmatic and causal values together, the possible consequences of CETI success range from the realization of our fondest fantasies to the actualization of our worst nightmares. The same is true of SETI success. ETIs could cook us for dinner whether or not we had succeeded in communicating with them—and while they can't convey new sciences to us without CETI success, they could still be the causes of wonderful effects. For example, they could inoculate us against all diseases even if they couldn't communicate with us, just as we inoculate dogs and cats.

But how likely are these fantasies and nightmares? When we evaluated the causal effects of ETI success and failure, we found reason to believe that the fantasy of universal disarmament and the nightmare of universal Weltschmerz were very low-probability consequences. In these cases, we had something to go on because the hypothesized consequences were hypotheses about how *we* would behave under conditions that were not entirely unlike conditions that we've faced in the past. The hypothesized consequences of SETI and CETI success, however, are hypotheses about how *aliens* would behave. Clearly, we have no more basis for an informed opinion about these matters than we have on the probability of ETI, SETI, or CETI success itself. The result is that the value of SETI or CETI success is itself radically indeterminate. It could range from the greatest imaginable boon to the greatest imaginable curse—from immortality to being cooked and eaten.

So what's the expected value of SETI or CETI? It's given by the formula

$$EV = p(s)v(s) + p(f)v(f),$$

where $p(s)$ and $p(f)$ are the probabilities of success and failure, and $v(s)$ and $v(f)$ are the values of success and failure. Earlier in this section I argued that the value of SETI or CETI failure is zero. Thus the expected value of SETI or CETI as a whole is equal to the product of the probability of success and the value of success. But both of these quantities are radically indeterminate. In section 4, we saw that $p(s)$ is vague over the entire interval from 0 to 1. In the present section, we saw that $v(s)$ can range from the highest good that we can conceive to the worst evil that we can conceive. The upshot is that the expected value of SETI or CETI itself is indeterminate between the highest good and the worst evil. Everything that we know is compatible with our receiving the gift of immortality with a probability of 1 if we engage in SETI, or with our being cooked and eaten with a probability of 1, or with anything in between.

Thus rationality provides us with no guidance for how to answer the meta-question about SETI or CETI. Spending a fortune on trying to make contact with extraterrestrials has as much rational warrant (namely, none) as spending a fortune on trying to *avoid* contact with extraterrestrials. Everybody's predilection is as good as everybody else's. To be sure, we've only considered pragmatic and causal values so far. But this null result isn't altered when we bring epistemic and gossip values into the picture. In the first place, it was argued in section 1.2 that the several extraterrestrial projects have not been shown to have a bearing on any theoretical issues. In this regard, the fact that there are (or aren't) any ETIs is very different from the fact that the perihelion of the planet Mercury revolves about the sun by an amount of 43 seconds of an arc every hundred years. The latter gives us an excellent reason to accept Einstein's paradigm-breaking views of space, time, and gravitation. The former doesn't change much of anything in our cognitive world. But this is to say that the extraterrestrial projects have no epistemic value to speak of. If tomorrow we received a radio message from space that was clearly of artificial origin, there would be huge headlines in the newspapers for a few days or weeks or months. But when the dust had cleared, none of our scientific ideas about the universe would have been changed. As things stand at present, the search for extraterrestrial intelligence is scientifically *boring*.

This conclusion is sometimes denied by biologists. For example, Mayr (1985) seems to suggest that evolutionary theory leads to the expectation that there are no ETIs. If this is right, then the discovery of ETIs would be an unexplained mystery calling for revisions in our received views about evolution. But evolutionary theory has nothing to do with Mayr's anti-ETI conclusion. Mayr's argument is purely inductive: of the billions of species that have existed on earth, only one has possessed sufficient intelligence to engage in SETI (see section 1.2). It isn't evolutionary theory that tells us that intelligent species are improbable. The force of Mayr's argument would be unchanged if, instead of believing in Darwinian evolution, we thought that all the species were created by special acts of God. The inductive argument would then be that ETI is improbable because the available evidence tells us that God hardly ever chooses to produce intelligent species. Establishing that there are (or aren't) ETIs neither confirms nor disconfirms the hypothesis of Darwinian evolution. It merely adds a detail to our natural history of the universe.

So bringing epistemic values into the picture doesn't alter the answer to the metaquestion. It only adds zeros to the zero value of minimal ETI success and to the enormous range of positive and negative pragmatic/causal values associated with SETI and CETI success. That leaves us with the gossip value of the outcomes of the several extraterrestrial projects. There's no doubt that the facts about ETI have a lot of gossip value—almost as much, perhaps, as the facts about the O. J. Simpson murder trial. This is reason enough for engaging in a minimal ETI project—a project of thinking about whether ETIs exist *without* trying to make contact with them. When it comes to SETI or CETI, however, the addition of a middling positive value to the maximally wide range of pragmatic and causal values once again leaves the indeterminacy unchanged. We couldn't decide whether SETI was pursuitworthy because SETI success might either make us immortal or get us eaten. Now we realize that SETI success might either (1) make us immortal and gratify our curiosity, or (2) get us eaten and gratify our curiosity. The value is not less indeterminate than before. The interval of values has merely shifted a notch toward the positive. So long as the positive gossip value of SETI success has a smaller absolute magnitude than the negative value of the destruction of all life on earth, the answer to the metaquestion will still be indeterminate.

Finally, it's worth mentioning that the answer to the SETI metaquestion would remain indeterminate even if we were to get a fix on the probability of SETI success. To the extent that SETI is likely to fail, the range of expected values for SETI diminishes. But the metaquestion remains unresolved because, for all $p(s)$ other than $p(s) = 0$, the range of associated values continues to straddle both sides of zero (i.e., the value of the status quo). As the probability of success diminishes, the stakes get smaller, but the decision whether to pursue SETI doesn't become any easier to make. With $p(s) = .0001$, we merely take a minuscule risk of being exterminated by aliens if we pursue SETI. But we also obtain a minuscule chance of being granted immortality. So there's no rational resolution to the metaquestion all the way down to $p(s) = 0$. When $p(s)$ is exactly zero, the metaquestion whether to pursue SETI disappears, for there's nothing left to pursue. We've already got our answer: SETI fails. So for all probabilities of success, the question whether to pursue SETI either has no answer or it doesn't arise. The same can be said of CETI.

Chapter 3

SETI Versus CETI

What is the relation between SETI and CETI? Does success at one of them entail success at the other, or does each one proceed to its own fate independently of the other? It may seem obvious that success at CETI is more demanding than success at SETI—more precisely, that (1) you *can't* succeed at CETI without prior or concomitant SETI success, but (2) you *can* succeed at SETI without succeeding at CETI. The first claim asserts that you can't talk to aliens unless you find some first. The second claim says that you might encounter intelligent extraterrestrials but be unable to learn their language. These propositions may sound incontrovertible, but the second one has been controverted. Donald Davidson (1974) has famously argued that the only way we can establish that an alien creature's vocalizations (or tentacle waggings) comprise a language is to succeed in developing a scheme for translating those vocalizations (or waggings) into sentences of our own language. But to establish that a creature has a language is to succeed at SETI, and to be able to translate that creature's utterances is to succeed at CETI. If Davidson is right, you *can't* have SETI success without concomitantly succeeding at CETI. The two projects succeed or fail together.

What are we to make of this claim? It's undoubtedly true that the construction of a successful translation manual—a manual that makes sense of the creature's present and future utterances—is excellent evidence for the hypothesis that these utterances are part of a language. But Davidson maintains that it's the *only* evidence for ETI that we can hope to obtain. I will discuss three strategies for deflecting the force of Davidson's claim.

The first strategy is based on an anti-Davidsonian argument that I've described elsewhere (Kukla, 2005). Here is a summary of my counterargument, altered slightly to fit its new context. Let's suppose that we come into contact with two extraterrestrial species, the Aldebaranians and the Vegans. Both these species have recognizable mouths and vocal chords, and both of them engage in a lot of vocalizing. We can't make any sense of Aldebaranian vocalization. Hence, Davidson would say, we can have no basis for believing that the sounds they make are speech acts. But we have managed to make sense of a lot of the Vegans' utterances. In fact, we've learned enough Vegan to be able to translate most English sentences into that language. In addition, the Vegans have shown themselves to be reliable informants: when their claims are investigated, they generally turn out to be true. Therefore, by everyday inductive reasoning, it's rational for us to believe their claims when we *can't* investigate them. Now for the last supposition: one of the things that the Vegans tell us is that the Aldebaranians have a language. How do they know? Because they've learned to speak it. It occurs to us that if they've learned Aldebaranian, they ought to be able to provide us with an Aldebaranian-Vegan translation manual. Since we speak passable Vegan, such a manual should provide us with the Rosetta stone that unlocks the Aldebaranian language. The Vegans admit that they have such manuals. However, they also tell us that these manuals are of no use to us humans. According to them, the set of Vegan sentences that have English equivalents has no members in common with the set of Vegan sentences that have Aldebaranian equivalents. Consequently, there are no English equivalents to any Aldebaranian sentences. But then there's no chance that Davidson's criterion for ascribing a language to the Aldebaranians can be met. Yet the reliable Vegans *tell* us that the Aldebaranians have a language. This interstellar scenario is logically and physically possible—we could encounter reliable Vegans who tell us just these things about the Aldebaranians. If we did have such an encounter, the testimony of the Vegans would provide us with strong evidence for the hypothesis that the Aldebaranians have a language. To be sure, this evidence is defeasible. But what isn't? The important point is that it would be rational for us to believe that the Aldebaranians have a language—*even though we're unable to translate it.*

The foregoing scenario is, I think, an adequate refutation of Davidson's claim that you can't attribute linguistic capacity to a being without being able to translate its language. But this particular refutation (there are two more to come) fails to establish that there can be SETI success without concomitant CETI success. It's true (within the parameters of the scenario) that we can learn that the Aldebaranians have a language without mastering that language—and it's true that this is a SETI success without a concomitant success at a *particular* CETI project. But it doesn't follow that it's a SETI success in the absence of *any* CETI success. In fact, the probative scenario specifies that there *has* been a CETI success: we've learned to communicate in Vegan! The scenario couldn't do its job if that important assumption were eliminated. Hence this scenario doesn't provide us with an example of SETI success without CETI success.

The second strategy for defusing Davidson's argument is to suppose that signals may have certain formal properties that identify them as moves in a language game, even if we have no inkling as to what the signals might mean. If we detected such a signal, we would know that there are language-using beings who are producing them, but we wouldn't be able to understand what they're saying. Thus we would experience a SETI success without a concomitant CETI success. Now it isn't obvious that there are any signal properties of the requisite kind. For this suggestion to be anything more than a tremulous hope, we would have to show that there are some properties that do the job. What might such a property be? Well, linguistic utterances are certainly going to exhibit nonrandom fluctuations in some dimension or other. So what about nonrandomness as an indicator of linguisticality? The problem, of course, is that there are countless nonlinguistic phenomena that produce nonrandom signal fluctuations. One interesting example is the emission of radiation from pulsars, which occurs with such clockwork regularity that it's been mistaken for alien communication by some overly eager CETI researchers. Nonrandomness of signal fluctuation is certainly a *necessary* condition for the signal to be linguistic. But it just as certainly isn't *sufficient,* and it's sufficiency that's required to do the job.

A favorite candidate in the CETI community for the job of language indicator is the series of prime numbers, 1, 2, 3, 5, 7, 11, . . . It's noted that there is no known physical process that implicates the property of primeness. Thus, if we were to receive a series of electromagnetic signals that recapitulated the series of primes, we would be justified in inferring that they were produced by an intelligent (i.e., language-using) agency. This inference is far from being apodictic. It's not as though we had any principled reasons for believing that primeness doesn't figure in natural phenomena. We believe it on the basis of a rough and ready induction: we've done a lot of science over the past few centuries, and we've never yet had to refer to primeness in the formulation of any law or generalization; therefore, we don't expect to need any reference to primeness in the future. Still, a rough and ready induction is something—it allows us to say that we may encounter evidence that favors SETI success without advancing us toward CETI success.

An important question for extraterrestrial studies is whether there are any signal properties that are guaranteed to be indicators of language use. We may be inclined to think that there aren't any such properties by the contemplation of parrots. These nonlinguistic creatures are capable of mimicking human speech so accurately that they would surely pass any purported test of linguisticality that doesn't involve any appeals to semantic notions like meaning or reference. But this avian capacity is parasitic on the existence of language-using beings for them to mimic. If we applied our test of linguisticality to a signal that was, unbeknownst to us, produced by parrots, we would conclude that there was a language-using agency behind its production—and we would be right. The agents wouldn't be the parrots, however. The question whether there are tests of linguisticality that are guaranteed to succeed is entirely open.

The third anti-Davidsonian strategy is to infer language possession from the presence of other accoutrements of intelligent life. This line is endorsed by Schick (1987):

> There are criteria other than translatability that we could appeal to to establish the existence of an alternate conceptual scheme. If creatures from another planet flew to the earth in sophisticated spaceships and proceeded to transform the planet by building complex structures, for example, then even if we could not translate their speech, we would still have good reason for believing that they possess an alternative conceptual scheme.

In this passage, Schick talks about the possession of an "alternative conceptual scheme" rather than a language. But Davidson explicitly equates conceptual schemes with languages. Thus, we're at liberty to substitute "language" for "conceptual scheme" in the quotation from Schick, whereupon it becomes a description of indirect evidence for SETI success. As in the case of primeness, the evidence to which Schick refers wouldn't be *conclusive,* for it would be amenable to the alternative explanation that the spaceships, planetary transformations, and complex structures are the products of innate technological expertise, like birds' nests and beaver dams. For all that, his example is good enough for the task it was designed to accomplish: the possibility of weak evidence is sufficient to block the Davidsonian hypothesis that you can have *no* other basis for attributing linguistic capacity to a being than the construction of a successful translation scheme.

The conclusion of this chapter vindicates our pre-analytic intuition: you *can* have SETI success without a concomitant CETI success. More precisely, you can be in possession of evidence that makes it rational to believe that you've encountered talking extraterrestrials, even though you understand nothing of their language.

Chapter 4

The One World,
One Science Argument

4.1 OWOS

Let's suppose that we come into contact with intelligent extraterrestrials. Will we be able to learn their language or to teach them ours? The project of establishing communication with an extraterrestrial intelligentsia—*CETI*, for short—faces some daunting obstacles. If our contact is mediated by radio signals, there's the problem of time: except for a handful of nearby stars, the time it would take for us to send a signal and get even an immediate a reply is going to be longer than the life expectancy of the sender. Figuring conservatively that learning a language requires a thousand exchanges, a course in Aldebaranian would take orders of magnitude longer to complete than the whole of human history. It's also hard to see how the connections between the terms of the language and features of the world could be conveyed without the capacity for pointing or otherwise calling attention to at least some of these features in the course of the language lesson.

But these are not the problems that I want to explore. Let's also suppose that we have unrestricted physical access to our intelligent extraterrestrials: when we feel like consorting with the Aldebaranians, we need only hop onto a faster-than-light ship that can get us to Aldebaran (or anywhere else in the galaxy) in 20 minutes. This assumption disarms the problem of time; it also provides us and the Aldebaranians with the opportunity to use props in the language lesson. Under these conditions, is there any difference between an Englishman's traveling to Aldebaran for an immersion course in modern Aldebaranian and his going to Japan to learn Japanese?

Well, at least we can expect a difference in degree. It's to be expected that the concepts enshrined in a language are going to be shaped by the geography and culture *(inter alia)* of the speakers. For example, terrestrial languages that evolved in tropical regions won't have a word for snow, and the languages of the arctic probably won't have a word for the activity of surfing. By the same token, the language of each culture is going to possess the means for referring to rituals and social conventions that exist in only that culture. Roughly speaking, the greater the differences in culture and geography, the greater the dissimilarity of lexicons, and the greater the difficulties in learning one another's language. Now English culture and English geography are very different from their Japanese counterparts—but they're not going to be nearly as different as the Aldebaranian environment. So learning Aldebaranian will probably be much more difficult than learning Japanese. It might even be beyond human cognitive capacities. Let's call this the *soft* argument against CETI.

The soft argument doesn't claim to warrant the conclusion that CETI *must* fail. It's a more-or-less kind of argument. The suggestion is that we have reason to expect that CETI is going to be an extraordinarily difficult enterprise—maybe too difficult to accomplish—but maybe not. CETI enthusiasts can still rationally hope that if enough time and effort is lavished on the program, success may be within their grasp. The most damaging conclusion that can be drawn from these preliminary considerations is that a successful outcome isn't going to be easy.

Moreover, Carl Sagan and other CETI advocates (let's call them *cetists*) have argued that there is cause for optimism. Aldebaranian culture and geography may be radically different from our own, but (it's argued) we're guaranteed to share at least one large topic for conversation: the laws of science and mathematics. These are purported to be the same everywhere in the universe, regardless of local variations in culture or planetary environment:

> [H]ow could we possibly decode such a[n extraterrestrial] message? [. . .] The message will be based upon commonalities between the transmitting and receiving civilizations. Those commonalities are [. . .] what we truly share in common—the universe around us, science and mathematics (Sagan, 1973, pp. 217–218).
>
> We have a lot in common [with intelligent extraterrestrials]. We have mathematics in common, and physics, and astronomy (Purcell in Cameron, 1963, p. 142).
>
> We may fail to enjoy their music, understand their poetry, or approve their ideals; but we can talk about matters of practical and scientific concerns (Anderson, 1963, p. 130).
>
> If physics is universal, they will make use of the same laws, although discoveries and applications associated with them will be diverse. [. . .] With a broad agreement over the principles of science, it is presumed that there would be few problems in constructing a dialogue based upon technology, science and mathematics (Lamb, 2001, p. 48).

Nicholas Rescher (1985, 1998) calls this the *one world, one science argument.* I'll refer to it as *OWOS.*

Let's expand OWOS. OW is the telegraphic version of the premise of the argument; OS is the conclusion. (Actually it's the penultimate conclusion—see section 4.10 below.) The import of OW is that the laws of nature are the same everywhere in the universe. E equals mc^2 in the vicinity of Aldebaran as well as on Earth. To the extent that Aldebaranian and terrestrial scientists do their work properly, their scientific literature will contain statements of those laws. Since the laws are the same, the content of their science will also be the same. That is to say, OS. Cetists regard this piece of reasoning as so self-evidently sound that it doesn't require any further defense. It's just a matter of calling attention to the obvious. But in fact, every point—the premise OW, the conclusion OS, and the conditional claim that OW entails OS—is beset by controversy. I will indicate the broad outlines of these controversies in sections 4.2 and 4.3; but these sections merely provide background material for my main point. I want to show that OWOS fails even if its contentious assumptions about science and the world are granted—that is to say, it fails on uncontentious grounds.

4.2 OWOS AND SOCIAL CONSTRUCTIVISM

One might have thought that at least the premise that we and the Aldebaranians live in the same world is uncontroversial. But it's denied by a substantial minority of philosophers of science and a clear majority of sociologists of science. According to these *social constructivists,* scientific laws (as well as particular matters of scientific fact) are *negotiated* rather than discovered (Latour & Woolgar, 1979). In this regard they're rather like *prescriptive* laws: what determines which ones are true is not an objective, mind-independent state of the world, but rather the state of play among socially certified experts (scientists, legislators, etc.) who strive to make their views prevail. Social constructivists don't merely make the relatively innocuous claim that the acceptance of scientific hypotheses is influenced by social forces. Latour and his colleagues maintain that scientific truths are *constituted* by social forces. E *is* equal to mc^2—it's not just a matter of belief—but the mass-energy equivalence is what it is *because* the party that championed relativity was victorious.

If you're a social constructivist, you're not going to be inclined to accept the premise OW—for there's no reason to suppose that the relevant negotiations on Aldebaran are going to have the same outcomes as the negotiations on Earth. It would be an incredible coincidence if they did. Thus if social constructivism is right, it's overwhelmingly likely that we and the Aldebaranians *don't* live in the same world in the relevant sense.

Are the constructivists right? This is arguably the most heated intellectual issue of the past 20 years. I don't propose to open yet another front of the science wars. Just for the record, my view is that constructivism is incoherent (see Kukla, 2000). By my lights, the OWOS argument is safe from the objection that its premise is false. Be that as it may, one would have expected cetists who deploy OWOS to address the issue of constructivism—at least to cite some criticism of the constructivist thesis. But none of them do.

4.3 OWOS AND CONCEPTUAL RELATIVISM

I turn now to the case for OS. One doesn't have to be a flagrant subjectivist to doubt that we and the Aldebaranians subscribe to the same scientific principles. For the sake of the argument, let's grant the premise OW—that is, that the laws of nature are the same everywhere in the universe. Here is one reason why OS doesn't follow: whether a true statement makes it into our compendium of scientific facts depends not only on the nature of the mind-independent world but also on the concepts and categories we employ for talking about the world. But (the argument goes) we're epistemically free to devise any concepts and categories we like—and so are the Aldebaranians. Thus there's no reason why we should expect the Aldebaranians to make the same conceptual choices as we do, even if we all live in the same world. Here's how Nicholas Rescher puts it:

> [. . .] the *conceptualization* of [extraterrestrial] science might be very different. We must reckon with the theoretical possibility that a remote civilization might operate with a radically different system of concepts in its cognitive dealings with nature. It is (or should be) clear that there is no single, unique, ideally adequate concept-framework for "describing the world." The botanist, horticulturist, landscape gardener, farmer, and painter will operate from diverse cognitive "points of view" to describe one selfsame vegetable garden in very different terms of reference. It is mere mythology to think that the "phenomena of nature" can lend themselves to only one correct style of descriptive and explanatory exposition. There is surely no "ideal scientific language" that has a privileged status for the characterization of reality. [. . .] Different cognitive perspectives are possible—no one of which is more adequate or more correct than any other, independently of the aims and purposes of their users (1985, p. 87).

In other words, the pie of reality is sliced by our conceptual systems into pieces of different sizes and shapes. The pie itself is a given—it's the one and only world the existence and uniqueness of which is affirmed in the anteced-ent of OWOS. But we're free to make slices of any kind that we like or find useful, and so are Aldebaranians. With this latitude, it would be extremely unlikely that the Aldebaranians' conceptual systems turned out to be the same as ours.

It's important to see that this criticism of OWOS fully grants the premise that the facts of science are the same everywhere in the universe. Let C1 and C2 be two conceptualizations of the same reality pie. Then there will be a true description P(C1) of the pie in terms of C1 and another true description P(C2) of the same pie in terms of C2. These two descriptions will not contradict each other. In fact, two statements *can't* contradict each other unless they utilize the *same* concepts (e.g., "Snow is white" and "Snow is not white"). Evidently, the world has these objective and universal properties—that it reveals the face P(C1) when it's interrogated with the conceptual scheme C1, and that it reveals the different face P(C2) when it's approached from the standpoint of C2.

In the passage quoted above, Rescher relies on a broad consensus among contemporary philosophers for *conceptual relativism*—the view that there is no privileged conceptual scheme. Rescher makes it sound as though the con-sensus is exceptionless and the issue is fully settled. But this overstates the case. It's true that most contemporary philosophers are conceptual relativists. There are, however, significant dissenting voices. The main one is Donald Davidson's (1974). Davidson claims that the crucial relativist's distinction between conceptual schemes and the unconceptualized world is incoherent. On this issue, I'm with the anti-cetists like Rescher—I think that Davidson's arguments don't work (see chapter 3). I won't argue the point here, however. The criticism I want to rely on for the repudiation of OWOS is discussed in the next six sections. Regardless of whether the coming argument succeeds or fails, cetists like Carl Sagan stand accused of not responding to their philo-sophical adversaries, the constructivists and the conceptual relativists.

4.4 OWOS AND THE SELECTION PROBLEM

Let's be generous to the cetists. Let's grant them that all intelligent beings utilize one and the same conceptual scheme. Then (if OW is also granted) the scientific principles available to be discovered are the same for us and for the Aldebaranians. But it's still a *non sequitur* to conclude that OS is true, that is, that the scientific principles *actually discovered and subscribed to* are the same for us and for the Aldebaranians. The problem for CETI arises from the fact that we—and presumably the Aldebaranians as well—don't know *all* of the available science. Humans and Aldebaranians will each possess a proper subset of the set of all the scientific truths that are expressible in their shared conceptual scheme. But then the part that we possess may not overlap with the part that they possess. I call this the *selection problem* for OWOS. Because of the selection problem, OS doesn't follow from OW even if it's assumed that everybody uses the same concepts. Moreover this refutation of OWOS doesn't depend on any challengeable philosophical assumptions. The only new premise required is that nobody knows every single scientific fact.

The selection problem merely points to a possibility. I haven't given the reader any reason to suppose that our science and the Aldebaranians' science *don't* overlap. I've claimed only that the assumption of OW leaves it open that the human and Aldebaranian scoops of true science *might* not overlap. Moreover, the arguments in the next six sections follow the same pattern: proposed solutions to the selection problem are rejected on the grounds that they're incompatible with certain speculative scenarios. This has impelled some readers of an earlier version to object that I attack the cetists' creed with mere logical possibilities. And so I do. But in this context, mere possibilities are enough to effect a refutation. Let me explain.

We've assumed that OW, the premise of the cetists' argument, is true. The quotations from Sagan and other cetists are essentially assertions that OW entails OS. The conclusion is OS. The entailment relation between OW and OS need not be deductive. The case for OS is a good one if the move from OW to OS is based on an acceptable ampliative inference. So my critics are right to claim that OWOS is not refuted merely by citing a logically possible state of affairs in which OW is true and OS is false. But this is only how the refutation starts: I describe a state of affairs X which, if true, would falsify OS. (For the argument in this section, X is the state of affairs wherein our scoop of scientific truth has no overlapping contents with the Aldebaranians'.) The existence of such an X shows that OW doesn't logically entail OS. But that isn't the point.

Let's see what can be concluded from the several epistemic relations we might have to X. To begin with, we might know X to be true. If we did, then we would also know that OS is false and OWOS would be shown to be irretrievably unsound. But of course we don't know that X is true. Alternatively, we might know that X is false. In that case, the mere logical possibility that X is true has no effect on the status of OWOS. What if we ascribe an intermediate degree of probability to X? We know that the conjunction OW & X entails ~OS. Thus

$$p(OW \ \& \ X) \leq p(\sim OS) = 1 - p(OS).$$

Equivalently,

$$p(OW)p(X) + p(OS) \leq 1.$$

But OW is a presupposition of the discussion. Thus p(OW) = 1, which means that

$$p(X) + p(OS) \leq 1.$$

This confirms our preceding conclusions: if X is known to be true (p(X) = 1), then OS is also known to be false (p(OS) = 0); but knowing that X is false (p(X) = 0) places no constraints on the status of OS (0 ≤ p(OS) ≤ 1). Moreover, if X is likely to be true (p(X) ≤ 0.5), then OS is likely to be false (p(OS) ≤ 0.5); and ascribing a low probability to X (p(X) ≤ 0.5) is compatible with ascribing either a high or a low probability to OS (p(OS) ≤ 1—p(X) and p(X) ≤ 0.5).

So we have an argument against OWOS if it's at least likely that two independent scoops of science don't overlap, and we *don't* have an argument against OWOS if that state of affairs is *un*likely. Confronted with a logically possible scenario that's incompatible with OS, cetists need only give us a good reason for supposing that it's unlikely that the scenario obtains. But what if no such reasons are forthcoming? A case in point is X, the hypothesis that two independent scoops of scientific truths won't overlap. I maintain that we currently have no good reasons for supposing that X is likely to be true, *and* we have no good reasons for supposing that X is likely to be false either. Until somebody presents a persuasive case for one side or the other, the indicated epistemic stance toward X is one of agnosticism. Translated into probabilistic terms, this means assigning a *vague* probability to X—that is, an *interval* of probabilities representing the state of opinion that the probability of X is one of the values in the interval, but that we're not saying which. For the vague probability to represent an agnostic state of opinion, the interval presumably has to span some values greater than 0.5 and some values less than 0.5. For the sake of definiteness, let's say that X is maximally vague—that its associated interval is the full range of probabilities [0, 1].

If the correct epistemic stance toward X is agnosticism, what is its import for OWOS? We saw above that if p(X) is low, then cetists need not rationally change their optimistic assessment of OW. But they can't take that line if they and their critics agree to agnosticism with respect to X. For suppose that they persist in maintaining the optimistic view that p(OS) = 0.8. Then, because p(X) + p(OS) ≤ 1, they also have to say that p(X) can't be greater than 0.2. But this is to depart from the agnosticism toward X that we've agreed is the correct stance. Consistency demands that we adopt an agnostic stance toward OS as well. But to do this is to admit that OWOS has failed. The cetist begins with an optimistic view of the prospects of OS. Then he encounters the idea of a state of affairs which is incompatible with OS and about the occurrence of which he's agnostic. The result is that he has to give up his optimism and

become agnostic with respect to OS as well. This is not as devastating a critique of OWOS as would be obtained by showing that an X is both incompatible with OS and *true*. The latter is an irremediable refutation, whereas the
former allows that future developments may give us a reason to ascribe a low
probability to X, thereby reinstating OS. But unless and until this development takes place, we may not rationally endorse OWOS. If we were OWOS
optimists before, the citation of an incompatible X about which we're agnostic is enough for OWOS to suffer an epistemic demotion.

How can you tell whether the probability of a hypothesis like X is vague
or simply low? If you're a Bayesian personalist, you don't tell—you *decide*
in any manner that you choose. Different people may make different decisions, as a result of which one and the same argument may be effective for
some recipients and not for others. An argument is significant to the extent
that it's effective for a lot of people. I believe that the selection argument
of this section is significant in this sense. As far as I know, nobody has ever
given us a reason to suppose that our selection of scientific truths is either
likely or unlikely to overlap with the Aldebaranians'. Until such reasons
are forthcoming, we have no choice but to regard the probability of this
hypothesis as very vague. As noted above, most of the coming arguments
have the same structure. Contra the comments of my early critics, these
arguments are not defused by the observation that they appeal to mere possibilities rather than actualities. But they can be refuted by giving a good
reason to suppose that the possibility is implausible rather than vague.

4.5   THE FUNDAMENTAL LAWS SOLUTION

Here's a proposed solution to the selection problem that I made up. Cetists
could try to argue that while the totality of available scientific truths may
be too voluminous to be completely mastered, this doesn't apply to the
fundamental laws of science—the brute truths about the universe that aren't
consequences of other truths. So while our common quest for scientific truth
doesn't yield a common lore, the quest for fundamental truth does.

One problem with this suggestion is that it's by no means obvious that the
best science does result in a small number of fundamental laws. There are at
least three alternative possibilities. First, the universe may be so complex that
there is a huge number—perhaps even an infinite number—of irreducible
brute truths. If this is so, there will be a selection problem for fundamental
laws that's just as problematic for OWOS as the general selection problem.
Second, there needn't be any fundamental laws at all. There may instead
be an infinite regress of laws, each one being explained by its predecessor.

Third, there may be many—perhaps infinitely many—equally adequate total theories of the universe, each one postulating a different set of fundamental laws. Moreover, even if there is a single law that explains everything and has no competitors, there's no reason to believe that any intelligent species has ever attained it or even come close. *We* certainly haven't attained it—so why should we suppose that anybody else has?

I have no idea how likely or unlikely it is that the universe can be uniquely described by a small number of fundamental laws. There's no question that this hypothesis is associated with a vague probability for *me*. If cetists take the same view, then they can't cite the fundamental laws argument as support for their continued adherence to OS. They have to give us a reason to suppose that the requisite conditions on fundamental laws are likely to be true. The ball is in the cetists' court.

4.6   THE MATHEMATICS SOLUTION

What about mathematics? Cetists are strongly inclined to the view that mathematics is universally appreciated among intelligent beings (e.g., Freudenthal, 1985). Of course, they hold the same view of physics. But the universality of mathematics is taken to be even more obvious (and even less in need of justification). Yet the reasoning that produces the selection problem for science applies equally to mathematics: even if it's conceded that all intelligent species possess *some* mathematics, it doesn't follow that any two of them have the *same* mathematics.

It might seem otherwise. It might seem that a species (or an individual) logically has to learn certain *elementary* mathematical truths before it can comprehend the more advanced results. Arithmetic comes before analysis. Thus if every intelligent species has *some* math, everyone will have the most elementary math—and the selection problem is solved. One obvious soft spot in this proposed solution is the premise that all intelligent species possess some mathematics. I will return to this assumption in section 4.9. Let's agree not to question it until then. Let's also grant that there are advanced branches of mathematics that can't be mastered without prior knowledge of other, more elementary branches. It's still a non sequitur to conclude that some species must have the same mathematics as some other species, or even that there's a reasonable chance that they do. For all we know, every species knows a different portion of elementary math. Once again, in order to avail herself of the elementary-math solution, the cetist has to give us a reason to suppose that this eventuality is improbable.

4.7   THE RADIO SOLUTION

Let's turn to a rather different kind of solution. Almost all of our attempts at communication with extraterrestrials have involved the sending and receiving (more accurately, sending and waiting around for the reception) of radio signals. Were a communicative link to be established in this way, we would be assured that our communicant possessed the technological means for constructing and operating radio transceivers. The claim has been made that sharing the machinery of radio communication ensures that we will possess a common scientific lore, namely those scientific facts that are necessary for building and using radios (Lamb, 2001, p. 33; McConnell, 2001, p. 195). But is it necessarily the case that radio can only be enjoyed by those who share our knowledge of the relevant science? David Raup argues that the right sort of adaptive pressure could result in organisms that perceive and emit radio signals without possessing intelligence at all:

> [. . .] there is at least one alternative route to sophisticated technology in bio-
> logical systems. It is the utterly simple, familiar process of adaptation and has
> no requirement of intelligence! [. . .] I am suggesting that any evolving bio-
> logical system has the potential of sending detectable evidence of its presence
> into space. You may argue that the sort of adaptive phenomenon I am describing
> would be unlikely to lead to the sending of radio or other signals into space.
> Why would an animal evolve an awareness of the space environment? What
> adaptational problem could be solved by producing the technology necessary
> to penetrate space? The answers to these questions are not straightforward but
> one aspect should be noted. Radio capability was developed by the human spe-
> cies solely for earthbound communication between members of the species. It
> was only later that the technology was converted to the investigation of space
> (radioastronomy). Also, most of the transmissions from the earth are inadvertent
> leakages from communications systems. So an interest in space *per se* is unim-
> portant and unnecessary (1985, p. 41).

Of course, Raup's space cows wouldn't provide us with an opportunity to test our linguistics skills. But if unintelligent organisms can have radio, it goes without saying that intelligent beings can also have radio without intelligently designing and constructing them. The intelligent Aldebaranians might rely on instinctive mechanisms for their radio communication, much as we intelligent humans rely on instinct for breathing or digesting our food. One can imagine intelligent fish who consciously work out the science for constructing devices that enable them to extract oxygen from the atmosphere. These beings could be said to breathe intelligently. The human method of radio communication would be akin to the piscine method of respiration, whereas Aldebaranian radio would be akin to human breathing. So the possession of radio doesn't ensure that the possessors have some of the same science as we do. Without an argument showing that the space cows scenario is unlikely to obtain, the possession of radio doesn't even make it reasonably likely that we share scientific lore with the possessors.

Another shortcoming of the radio solution to the selection problem is that it begs the question. Grant that the use of radio requires that the user possesses certain scientific facts. Obviously the needed facts are in *our* possession. But why should we believe that the relevant information is included in any *other* intelligent species' scoop of scientific truth? Maybe we're the only species in the universe that uses radio. To solve the selection problem, you have to give a reason for supposing that our science and extraterrestrial science are going to overlap. The reason being evaluated is that users of radio will have overlapping sciences. One could as well say that the selection problem is solved by the fact that believers in $E = mc^2$ have a science that overlaps with ours (the overlap being $E = mc^2$). The question, of course, is whether there's a reasonable chance that any intelligent extraterrestrials *are* believers in $E = mc^2$. Like believers in $E = mc^2$, users of radio also have a science that overlaps with ours (this has been granted for the sake of the argument). But the question again is whether it's at all likely that any extraterrestrials *do* use radio. To suppose that they do isn't to propose a solution to the selection problem—it's to suppose that there is no problem.

4.8   THE COMMON CONDITIONS SOLUTION

Consider the *evolutionary* hypothesis that adaptive pressures can inculcate certain innate ideas about the world—or perhaps that they instill a propensity to adopt certain ideas, or to abduce and evaluate them—these distinctions won't matter in the present context. It might be argued that these innate ideas will tend to be true: if green berries are poisonous, it's more adaptive to have the prejudicial belief that green berries are poisonous than to have the prejudicial belief that green berries are *not* poisonous. So it makes sense to suppose that our innate ideas about the world will find their way into the collection of our *scientific* ideas about the world: they will tend to present themselves to us early on as very promising hypotheses, and when put to the test, they will in fact tend to be confirmed. I don't endorse the evolutionary hypothesis. I'm just going to try to ascertain whether it can bring us any closer to a solution to the selection problem.

The evolutionary hypothesis specifies that the truths that end up in our repertoire of innate ideas (and so, probably, also in our scientific lore) are those that give us an adaptive advantage. Here is another hypothesis, which I call the *common conditions* hypothesis: there are conditions for survival that can be expected to obtain wherever intelligent life evolves. I don't endorse this hypothesis either. But it could be argued that if both hypotheses are true, then our innate repertoire (hence also our science) can be expected to overlap with the Aldebaranians'. The argument would go as follows. The common conditions hypothesis tells us that we and the Aldebaranians quite likely face some of the same problems of survival. Common problems yield to a common solution. Hence there are truths—those that provide the common solutions to the common problems—that would assist the survival of both Aldebaranians and humans. The evolutionary hypothesis tells us that these truths are likely to be innately accepted by both species. Therefore they're likely to pop up in both human science and Aldebaranian science. And therefore the selection problem is solved—if the evolutionary hypothesis and the common conditions hypothesis are true.

Most of the work in this argument is done by the common conditions hypothesis. In fact, one might try to run the argument on the basis of that hypothesis alone, dispensing with the evolutionary twist. One could claim directly that common problems of survival lead to common solutions, regardless of whether the solutions are hard-wired by evolution. If electromagnetic disturbance is a serious threat to their well-being, intelligent extraterrestrials will probably find out the truth about electromagnetism. So if most intelligent extraterrestrials are threatened by electromagnetism, many of them are going to find out the truth about electromagnetism—and so the selection problem will be solved. The two arguments—the one that utilizes the evolutionary hypothesis and the one that doesn't—engender pretty much the same dialectic. But they are different stories in detail. I address—and reject—the story that employs the evolutionary hypothesis. Since that scenario allows the cetists more room for defensive philosophical manoeuvres, its refutation automatically entails the refutation of its more restrictive non-evolutionary alternative.

It's a presupposition of the discussion that the evolutionary hypothesis is true. Should we accept the common conditions hypothesis? I don't know of any *theoretical* arguments derived from evolutionary theory to the effect that there is an item of knowledge that would be useful wherever intelligence evolves. If we had already encountered a dozen or so intelligent extraterrestrial species, we might be able to solve the problem by purely empirical means: we could look to see if there are any commonalities in the conditions for their survival. But we have only the single case of humans to go on. This is, of course, an inadequate basis for an inductive generalization—the fact that an aversion to green berries is advantageous to human survival doesn't provide any evidence for the hypothesis that green-berry aversion is a good thing for Aldebaranians. Once again, the ball is in the cetists' court—it's up to them to give us a reason to suppose that the common conditions hypothesis is true.

Moreover, the argument from the evolutionary and common conditions hypotheses to the likely existence of overlapping sciences doesn't go through. Let's assume that the common conditions hypothesis *does* turn out to be true. Suppose that the survival of every intelligent species is enormously helped by a knowledge of f = ma (though it's hard to see what use this knowledge can be to intelligent plants whose only interaction with the world is through olfaction). It still wouldn't follow that adaptive processes have furnished all or even some species with innate knowledge of that useful fact. For one thing, the threat to survival that's diminished by a knowledge of f = ma might also be reduced just as efficaciously by other knowledge. Perhaps there's an electrical solution to the problem of mechanics that's solved by f = ma. If, as seems plausible, there are many different ways to solve one and the same survival problem, then the assumption that the knowledge relevant to such a solution is implanted in our genes is not yet strong enough to underwrite the conclusion OS.

In fact, there's bound to be an alternative *non-cognitive* solution to any adaptive problem that can be solved by the application of knowledge. Suppose that an adaptive problem is solvable by the evolution of an innate knowledge of some fact F. Then it's also solvable by the evolution of the non-cognitive behavioral disposition to *act like a being that knows F*. This is illustrated in the discussion of radio in the previous section. In that section, it's pointed out that a species could develop the capability to make and use radios without relying on any propositional knowledge. The construction of radio transceivers could be guided by non-cognitive biological mechanisms, like avian nest-building and the construction of beaver dams. Alternatively, radio could be a built-in feature of the alien's physiology, like eyes and teeth. The same alternatives are available for any adaptive knowledge. In fact, it's reasonable to regard the behavior of terrestrial animals—the hops of kangaroos and gallops of horses—as guided by non-cognitive mechanisms that mimic the possession of substantial bodies of truths belonging to mathematics and the science of mechanics. There's a pickwickian sense in which a galloping horse is busy solving differential equations. Evolution *could* have solved the relevant problems of horse and kangaroo adaptation by literally making them natural-born solvers of differential equations; but instead it chose the non-cognitive route. However many conditions for survival we may share with the Aldebaranians, evolution could have solved *these* problems non-cognitively as well. Common conditions for survival don't assure common sciences. For all we know, they don't even make common sciences a reasonable bet.

4.9 THE INTRACTABILITY OF THE SELECTION PROBLEM

My analysis has proceeded by considering and rejecting a series of proposed solutions to the selection problem. This leaves it open that cetists might come up with new and unrefuted proposals. In this section, however, I present an argument to the effect that *there cannot be* a solution to the selection problem. This is a generalization of several of the previous arguments. We saw that intelligent beings don't have to know the fundamental laws of science or how to construct radios. In fact, *they don't have to know any science or mathematics at all.* The selection problem arises because two scoops of science may turn out to be non-overlapping. The new argument is that some scoops may be entirely devoid of science or math. These null scoops are *guaranteed* to have no overlaps with any other scoops—the intersection of the empty set with any other set is the empty set. Thus we have a recipe for refuting any candidate-solution to the selection problem. Suppose it's suggested that some scientific fact F occurs in every scoop. Whatever F may be, we can be sure that the anti-cetist will have this retort available: not in the null scoop! For all we know, every intelligent species except for our own selects the null scoop.

Cetists will, of course, deny the last statement. They seem to presuppose that there's a good chance that any intelligent beings are going to possess some kind of scientific lore. This is a gratuitous assumption at best. The boundaries of what counts as intelligence and what counts as science are admittedly vague. But if "science" is construed narrowly enough to have arguably universally valid principles, then *human beings* didn't have science until very recently. In fact, many human cultures never did develop science. Yet we humans are (by our own reckoning) intelligent if any beings are. The objection that you can't survive without some scientific lore—for example, the knowledge that if you step off a high cliff you'll fall to your death—is confuted by the example of dogs and cats. These creatures don't have any difficulty dealing with cliffs. Perhaps there's a sense in which they may be said to know the relevant facts about cliffs, but this putative knowledge certainly doesn't qualify as scientific.

The same point can be made without reference to cultural history. It's enough to note that there are people in our own era and society who possess exquisite verbal intelligence, but are stymied by the simplest high school algebra problem. For all we know, the planets of Aldebaran are populated exclusively by such humanities types. At present, we have no way of knowing that scientific intelligence isn't vanishingly rare, even among the galactic intelligentsia. Maybe the galaxy is teeming with intelligent beings, but they all devote their intellectual energy to writing poetry and perfecting their stand-up comedy routines.

4.10 THE SUPERFLUITY OF OWOS

OS isn't intended to be the ultimate conclusion to be derived by cetists from OW. Cetists like Sagan are interested in establishing the CETI hypothesis that communication with extraterrestrials is feasible. But OS makes no reference to communication. OS merely tells us that extraterrestrial science will overlap with ours. So there's an additional step in the argument that needs to be evaluated: the move from OS to CETI.

As Lamb writes in a previously quoted passage, "With a broad agreement over the principles of science, it is presumed [by cetists] that there would be few problems in constructing a dialogue based upon technology, science and mathematics" (Lamb, 2001, p. 48). That is to say, cetists *presume* that OS entails CETI; but they make no attempt to show, or even to sketch, how the possession of a common science can be parlayed into a knowledge of each other's scientific language. The presumed fact that we and the Aldebaranians accept exactly the same scientific facts ensures that there will be an English translation for every Aldebaranian scientific statement and vice versa. But it doesn't automatically follow that we'll be able to pair them up properly. Once again, the cetists need to tell us more.

Even if OWOS is sound, the most it can accomplish by itself is to refute the *soft* argument that CETI is unlikely because we and the aliens would not share any common topics of discussion. This leaves it open that CETI may be extremely unlikely or even impossible on account of other reasons. But as a mere refutation of the soft argument, it's superfluous—for it's obvious that one being can learn another's language without special difficulty even if they share no common topics of discussion. Babies do it all the time: they learn the language of beings with whom they share no prior topics of discussion, and they do it in a few years' time without expending any extraordinary effort. This isn't to say that the procedures followed by babies for learning English are also available to Aldebaranians. But if they're not available, it isn't because of the lack of common topics with the language teachers. So OWOS isn't needed to refute the soft argument.

If OWOS is nothing more than a (purported) refutation of the soft argument, then my critique of OWOS is nothing more than a refutation of a refutation. What I have to say doesn't by itself entail that CETI is impossible or unlikely. There may be reasons for optimism about CETI. I've just shown that OWOS isn't one of them. In fact, I think that there's a powerful reason for being pessimistic about CETI. That's what I'm going to discuss in the rest of the book. The coming critique won't depend in any way on the repudiation of OWOS—the objection I'm going to mount holds even if all intelligent species in the universe possess exactly the same science.

It will be useful to review all that's been granted. The following list includes some items that haven't been discussed before. For the rest of the book, it's assumed that:

(1) the universe is teeming with intelligent extraterrestrials;
(2) we have immediate and unlimited access to them;
(3) they want us to learn their language;
(4) they can indicate to us whether any particular thing belongs or doesn't belong to the extension of a given term in their language;
(5) for our part, we are prepared to spend unlimited time, effort, and money on the project of learning their language;
(6) the laws of nature are the same everywhere in the universe;
(7) all intelligent species practice science;
(8) they all employ the same conceptual scheme;
(9) they all have identical scientific knowledge;
(10) there's an English translation for every sentence in every extraterrestrial language, and vice versa.

Cetists can't complain that we've stacked the deck against them! Despite this philosophical largesse, the project of learning to speak Aldebaranian may very well come up against an insuperable obstacle. The obstacle has to do with Noam Chomsky's theory of language acquisition.

Chapter 5

Innateness, Linguistic Universals, and Extraterrestrial Communication

5.1 INTRODUCTION

From the late 1960s to the early 1980s, Noam Chomsky and Hilary Putnam, arguably the two main progenitors of the cognitive science revolution, engaged in a protracted, acerbic, and intellectually exhilarating exchange on the status of Chomsky's nativist account of language acquisition. Many of the arguments that were presented in this debate have become a part of the standard accounts of the nativist and empiricist stances. In this chapter, I will examine a clever—and very brief—argument of Putnam's that has not received the attention it deserves.

The subject is linguistic universals—grammatical and phonological features shared by all natural languages. I will use "LU" to designate both linguistic universals and the hypothesis that there are linguistic universals. The truth of LU is regarded as uncontroversial by both Chomsky and Putnam. It's also a presupposition of my discussion.[8] Chomsky (1967) had noted that one of the theoretical benefits of his brand of nativism was that it provided the best available theoretical explanation for this phenomenon. Putnam's (1967/1980) brief and clever argument was intended to be a refutation of this claim. He purported to show that the "common origin" hypothesis—the hypothesis that all natural languages are descendants of a common ancestral language—provides a *better* theoretical explanation than Chomsky's for LU. Chomsky (1969/1981) wrote a correspondingly brief reply that missed the point. Thirty years later, Putnam's treatment of LU received a (brief) notice from each side of the philosophical fence. A pro-Chomskyan article essentially repeated Chomsky's inadequate reply (Laurence & Margolis, 2001, p. 248); and an anti-Chomskyan book treated Putnam's hypothesis as a "speculation" rather than as the conclusion of a reasoned argument (Cowie, 1999, p. 180).

I will maintain that Putnam's argument ultimately fails, although not for the reasons cited by Chomsky and his more contemporary surrogates. Moreover, even if it had succeeded, its conclusion would not have constituted a devastating critique of Chomskyan nativism: Chomsky could have assimilated Putnam's claims about LU with minimal theoretical changes. Nevertheless, Putnam's argument deserves close scrutiny. Despite the fact that it fails to achieve its aim, it's a paragon of interdisciplinary arguments in cognitive science. It blends empirical and *a priori* considerations into a seamless unity—for example, it alludes both to universally acknowledged linguistic data and to philosophical principles of theory choice to arrive at its conclusion. It also derives strong and surprising consequences from apparently modest resources, and it does so in a few deft strokes.

More importantly, the ultimate failure of the argument doesn't necessarily mean that it's wrong on every point leading up to the grand conclusion—there may be one or more unrefuted subarguments embedded within a demonstrably unsound overarching structure. I will claim that this is the case with Putnam's argument. In the course of constructing his critique of Chomsky, Putnam makes use of a lemma that remains unrefuted, even if both my criticism of the broader argument and Chomsky's are fully accepted.

To say that the lemma is unrefuted is not yet to say that it's been firmly established. The few deft strokes need filling in. This need for expansion turns out to be another reason that Putnam's neglected argument is worthy of study: when carefully examined, the case for Putnam's lemma is found to depend on the status of a multitude of broad and significant hypotheses belonging to diverse fields of research. In addition to linguistics itself, the implicated fields include philosophy of science, probability theory, evolutionary biology, psychology, and history. Drawing out the import of Putnam's argument turns out to be an intellectual feast.

Having been apprised of the dishes to be served, the discriminating consumer might object that partaking in this particular feast would be an act of gourmandise. I am proposing to explore the import of a lemma used in an argument that I regard as unsound. But lemmas are not significant or interesting in and of themselves—they're merely way stations in the journey toward a significant conclusion. If the conclusion is given up, its associated lemmas lose their *raisons d'etre*. Getting clear on the status of an orphaned lemma is like reading yesterday's newspaper. This objection is reason enough to give up our concern for the general run of orphaned lemmas. But Putnam's lemma is an exception to the rule. The proposition that I call "Putnam's lemma" turns out to be used as a way station to *two* significant conclusions. The first is

Putnam's own contention that the common origin hypothesis provides a better explanation of LU than Chomsky' theory does. The argument from Putnam's lemma to this conclusion fails—or so I maintain below. But this failure leaves the argument from the lemma to the *second* significant conclusion untouched. The second significant conclusion is that we will never be able to converse with extraterrestrials—the dreams of Carl Sagan and other devotees of SETI will never come to pass. It will be seen that the argument from Putnam's lemma to the repudiation of extraterrestrial communication is transparently valid. All the work is in establishing the lemma itself.

5.2 THE INNATENESS HYPOTHESIS

Let's begin with a brief review of Chomsky's theory of language acquisition. According to Chomsky, first-language learners possess a "language acquisition device" that receives and processes linguistic data. These data are the statements that the learner hears spoken about her.[9] The language acquisition device works to formulate a set of syntactic rules from which the linguistic data can be reconstructed. When new data are encountered that can't be generated by the current rules, these rules are appropriately revised. When further revision proves to be no longer necessary, the learner has mastered the language. The final set of rules constitutes the *grammar* of the language.

The fundamental problem of theoretical linguistics, as Chomsky sees it, is to specify the properties that a device must possess in order to be able to work out the grammar of the language from the linguistic data. This cognitive task has nearly the same form as the theoretical physicist's task of fashioning a theory that accounts for a certain class of observational data. If we were asked to describe the workings of the "physical-theory acquisition device," we would tell a story involving logical deduction, enumerative induction, inferences to the best explanation, appeals to principles of parsimony, and indefinitely many other principles and inferential practices, some of which would elicit heated dispute. What unifies this collection of principles and practices is our opinion that it's *rational* to believe the principles and to engage in the practices. At present, we have very little to say about how rationality is (or isn't) implemented in our cognitive apparatus. So we can't even come close to providing a complete naturalistic description of the workings of the physical-theory acquisition device. But, though it doesn't take us very far, we can at least say that the device works on the basis of rational principles. This rules out making theoretical choices on the basis of a coin flip. Physics doesn't work that way.

But aren't the same things true of the *language* acquisition device? Doesn't it employ rational inferences in going from the linguistic data to their grammar? In fact, aren't the rational inferences employed by the language acquisition device the same as those employed by the physical-theory acquisition device: deduction, induction, and so on? In sum, don't both devices operate in the same way on their inputs? If this is so, then what's the point of distinguishing the two devices from each other? We might as well say that there's one general-purpose device that accepts both linguistic data and physical data as inputs and processes them in the same way to produce an appropriate theory.

We come now to the interesting point. Chomsky claims that the linguistic data available to the child don't provide enough information to effect a rational reconstruction of the grammar. This isn't due to any cognitive deficiencies of the child. In fact, it's Chomsky's contention that *no being, however intelligent, could parlay the linguistic data that are available into knowledge of the grammar by a process of rational inference.* The data are too "impoverished" to allow for the discovery of the complex and subtle rules of grammar:

> [T]he basic problem is that our knowledge [of grammar] is richly articulated and shared with others from the same speech community, whereas the data available are much too impoverished to determine it by any general procedure of induction (Chomsky 1986, p. 55).

The case for this thesis is known as the *poverty of the stimulus* argument. According to this argument, trying to work out the grammar of the language on the basis of a list of the sentences that we've encountered is akin to trying to discover the true laws of mechanics (i.e., relativistic mechanics) when the only data available to us are about slow-moving objects. The data are insufficient for making the desired inferential journey. You can't get there from here.

But then how can we account for the fact that everybody learns to speak? For that matter, if the data are indeed inadequate, how is it that *anybody* learns to speak? Chomsky's explanation for this remarkable fact is that we're all innately endowed with linguistic information. In effect, we're able to learn the grammar on the basis of inadequate linguistic data because the missing information is supplied from within. The data alone aren't enough to enable us to reason our way to the right grammar, but the data *plus* the innate linguistic information are jointly rich enough to do the job. Putnam (1967) has dubbed this the *innateness hypothesis* (IH). The poverty of the stimulus argument rules out a class of explanations for the phenomenon of first-language acquisition, and the innateness hypothesis provides us with an alternative candidate.

In brief, the Chomskyan remedy for the poverty of the stimulus dilemma is innate knowledge. Chomskyans refer just as often to "innate constraints." It's easy to see why this phrase has gained favor. If we approached language learning without any predispositions, we would have to consider all logically possible grammars as equally viable candidates for the correct grammar of the language we're trying to learn. The linguistic data enable us to eliminate some of these candidates. That is to say, the data function as *constraints* that would-be grammars of our language have to satisfy if they are to continue to be viable candidates. But these constraints still leave too many candidates in the running (any number more than one is too many). This is the conclusion of the poverty of the stimulus argument. IH asserts that there are innately *additional* constraints beyond those imposed by the data. Since these innate constraints perform the function of eliminating candidate grammars, they're roughly equivalent to the innate knowledge that the grammar is *not* one of the eliminated candidates. In this way, talk of innate knowledge is interchangeable with talk of innate constraints.

But these remarks are merely programmatic. Granting that there are innate constraints that whittle down the number of candidate grammars, how is this whittling effected? By what process is a grammar eliminated? There are various *prima facie* possibilities. The most literal construal of "innate knowledge" would be one that took the resemblance between the first-language learner and the theoretical physicist as far as it can go. On this account, the child is a little scientist who consciously and wilfully examines whether various hypothetical grammars are compatible with the data. The only difference with real science (if it is a difference!) is that there are some additional propositions that play the same constraining role as the data, but are not themselves data. The child groundlessly, dogmatically treats them as true.

One variation of this scenario is that everything happens as described above, except that it happens unconsciously. There are also alternative procedures that can have the same constraining effect as the dogmatic acceptance (whether conscious or unconscious) of certain propositions. For example, our cognitive apparatus might be so constructed that certain grammatical hypotheses are never generated and offered up for evaluation in the first place. One difference between this style of constraint and the dogmatic-knowledge style is that only the latter requires that the innate knowledge be explicitly represented in the learner's mind. I bring up the issue only to make the point that the choice of constraint styles has no bearing on the topics to be discussed in this chapter. I find it convenient to write as though we were innately supplied with internal representations of linguistic information. But I'm not endorsing this or any other constraint style. What I have to say below can be said just as well, *mutatis mutandis*, about any constraint style.[10]

5.3   THE WEAK ARGUMENT AND THE STRONG ARGUMENT

A confession: I've been intentionally sustaining an equivocation in my account of Chomsky's argument for innateness. The claim that children have inadequate linguistic data for learning their language by rational means can be interpreted in either of two ways. The *weak* sense of the claim is that children *as a matter of fact don't* receive enough linguistic data for a rational reconstruction of the grammar. The *strong* sense is that children *can't even in principle* obtain enough data to reconstruct the grammar. Chomsky himself almost always relies on the weaker claim. For example, his frequent references to "impoverished data" suggest that the story might have been different if the data were sufficiently enriched. Similarly, after discussing certain intricate transformations in English, Chomsky remarks:

> In this case, too, it can hardly be maintained that children learning English receive specific instruction about these matters, or even that they are provided with relevant experience that informs them that they should not make the obvious inductive generalizations (1980a, p. 4).

Placing the emphasis on the lack of specific instructions once again suggests that the instructions *might* have been provided. Chomsky also frequently talks of the *facility* with which children learn language as evidence for the innateness hypothesis.[11] Here again, the reference to facility presupposes the weak argument. If the argument were that it's *impossible* to learn a language by purely rational means, it would be pointless to inquire into whether children learn it easily or with great difficulty. The fact that they learned it at all would be the only theoretically relevant consideration. In relation to the weak argument, however, the facility of language learning is yet another manifestation of the general phenomenon of children's linguistic accomplishments outstripping the resources made available to them. In these and countless other places, Chomsky relies on the argument that the child just happens to receive inadequate data. We need innate constraints to make up for the deficiencies in the information presented to us by our harried parents and teachers.

It shouldn't be supposed that the weak argument is deficient in some way. That's not the import of calling it "weak." It's weak only in the sense that it presumes less. To say that language can't rationally be reconstructed on the basis of the available data is to say less than that language can't rationally be reconstructed on the basis of any data that *might* be made available. In fact, weak arguments are *better* at securing their conclusions than strong arguments: since they presume less, there's less that can go wrong with them. If its premise holds, the weak argument does everything for the innateness hypothesis that the strong argument could do.

So the logic of the weak argument runs as follows. Children learn their language on the basis of linguistic data which are demonstrably inadequate for the task. Therefore they must bring innate information to the task. For *my* purpose—not Chomsky's—it's important to note that this argument is entirely compatible with the view that language can also be learned without appealing to innate information. If the problem is inadequate data, one solution is to supply the missing data. The appeal to innate information undoubtedly makes the task of language learning easier. But there's nothing in the weak argument to suggest that we couldn't do without innate constraints if we had to. Of course, there's no way of knowing beforehand just how difficult it would be to reconstruct a language by a "general procedure of induction." Maybe imparting the data needed would take hundreds of years of regimented instruction. To have an informed opinion on this issue, we'd have to look in detail at what the deficiencies in the data are. Whatever they are, however, these deficiencies are in principle *remediable.*

The question of Chomsky's relation to the strong argument is complicated. I'll get back to this question in due time. In the meantime, I'll follow Fodor's (1975) fully explicit and unabashedly enthusiastic treatment of the strong argument. In Fodor's hand, the problem that Chomsky has grappled with in linguistics becomes a special case of a more general difficulty. In fact, the problem is none other than the most famous conundrum of twentieth-century analytic philosophy: Goodman's (1954) "new riddle of induction." The riddle is that there are infinitely many hypotheses, making infinitely many divergent predictions that are compatible with any finite set of data. Consider the data that all the emeralds that have been observed to date have been green. One hypothesis which is compatible with these data is the theory that all emeralds are green. Another is the theory that all emeralds are *grue*, where "grue" is defined as "green if first observed before the year 2100, and blue otherwise." Still another theory is that all emeralds are *gred* (green if first observed before 2100, red otherwise). And so on. The first theory leads to the prediction that the first emerald to be newly observed after December 31, 2099, will be green; the second theory predicts that this selfsame emerald will be blue; the third that it will be red; and so on. Obviously, no amount of additional data is going to make a difference. Nor will the overabundance of available hypotheses be diminished by the expedient of waiting until January 1, 2100. When the twenty-second century dawns, a quick glance at a previously unobserved emerald will immediately knock at least two of the three hypotheses mentioned above—that emeralds are green, that they are grue, and that they are gred—out of contention. But there will still be infinitely many hypotheses compatible with all the data. For suppose that it turns out that all the emeralds observed in the twenty-second century are green. One theory which is compatible with these observations is that all emeralds are green. Another is that all emeralds are grue-22, where "grue-22" is defined as "green if first observed before the year 2200, and blue otherwise." And so on. As long as the data in our possession are finite—and we may be sure that they always will be finite—there will always be infinitely many mutually incompatible hypotheses consistent with the data.

The difficulty is entirely general: one can construct gruish alternatives to any hypothesis about anything. In fact, we can gruify all of our scientific beliefs in one fell swoop as follows. Let S be all the laws and generalizations we currently believe to be true about the world. And let T be any alternative set of laws and generalizations, however bizarre and implausible they may be. T might entail that quasars are made of marshmallow, and that every time somebody utters the word "Manitoba," somewhere an elf splits cleanly in two, and so on. The only requirement is internal consistency. Then consider the following theory of the universe. Until tomorrow, the empirical consequences of S will have been true, but from tomorrow onward, the empirical consequences of T will be true. This theory of the universe has exactly the same relation to the data at hand as our current theory S does. So if we prefer S, there has to be some reason other than conformity to the data.

Formally, the relation between linguistic data and the grammar of the language is the same as between evidence and theory. It thus goes without saying that there are also infinitely many competing grammars that can account for any finite set of linguistic data. Suppose, then, that language learners select randomly from the set of grammars that are compatible with the data in their possession. In that case, the probability of their hitting on the *correct* grammar is 1 over infinity, or zero. The fact that children do learn their language is conclusive evidence that they do *not* search at random through the solution space of possible grammars. But what could be the basis for a non-random selection of hypothetical grammars? As before, the innateness hypothesis provides the needed explanation for the successful performance of language learners. To be sure, the innate information can't simply have the form of additional data, for no finite amount of data is sufficient to pick out a single grammar. But there are undoubtedly packages of information that would do the job. For example, our innate endowment might specify a finite number of possible grammars for the languages that we'll be called upon to learn. With only a finite number of alternatives to choose from, we may indeed hope that the finite amount of linguistic data available to us will be adequate for making a selection.

Lastly, and leastly, we turn to the question of Chomsky's personal relation to the strong argument. For the most part, Chomsky relies explicitly on the weak argument. This by itself tells us nothing about how he views the strong argument—for the two aren't incompatible. In fact, the validity of the strong argument entails the validity of the weak: if children *can't* get enough linguistic data in principle to reconstruct the language, then they certainly *won't* get it. In such a case, one might very well choose to make one's case on the basis of the weaker argument simply for the sake of elegance—there's no point swatting a fly with a sledgehammer. On the other hand, one can logically accept the weak argument and reject the strong. There is, however, at least one place where Chomsky endorses the strong argument outright:

> In the history of modern philosophy there is a vast literature dealing with a few very simple points about the impossibility of induction, like the whole debate about Goodman's paradox. Once you understand the paradox, it is obvious that *you have to have a set of prejudices in advance for induction to take place* (Chomsky & Fodor, 1980, p. 259).

It's perhaps significant that this passage comes from a joint symposium presentation with Fodor. Nevertheless, this sketch of the strong argument appears under Chomsky's name.

5.4   THE MODULARITY HYPOTHESIS

What's Putnam's objection to IH? It isn't a blanket aversion to innateness. He wouldn't, for example, object to the claim that we're innately endowed with the capacity to make valid inferences in accordance with modus ponens. In fact, he readily concedes that learning a first language, like learning anything else, requires an endowment of innate knowledge:

> *To be sure,* human "innate intellectual equipment" is relevant to language learning; ... But what rank Behaviorist is supposed to have ever denied *this*? (Putnam, 1967/1980, p. 244).

However, he thinks that an innate knowledge of broad, universal principles like modus ponens and inductive logic is enough to get the job done. As Putnamians are wont to say, all that's needed for first-language learning is a set of "general learning strategies." Chomsky, on the other hand, thinks that the linguistic data are so impoverished that the task of first-language learning requires a supplement of narrowly targeted, specifically linguistic information—rules about noun clauses, auxiliary verbs, and so on. Innate knowledge of this kind is utterly irrelevant to the enterprise of physics—the concept of a noun never makes an appearance in the development of physical theory. In contrast, knowledge of modus ponens is a prerequisite for success at virtually any cognitive task.

I noted above that when two cognitive devices have the same innate endowment, they are better conceptualized as one and the same device with two
kinds of inputs. Just as compellingly, if two cognitive tasks require drastically
different innate endowments, they are best thought of as being done by different devices. Thus if Chomsky is right, language acquisition is the province
of one cognitive device among several. Each of these devices, or *modules*,
bears a distinct innate endowment that enables it to handle a delimited
class of tasks. Different modules may also utilize different constraint styles.
According to Chomsky, language is learned by the operation of a language
module. Whether physics is done by a physics module depends on whether
the conduct of physics relies on an innate knowledge of physical information.
The alternative is that the general learning strategies that are implicated in all
cognitive tasks are enough to account for our performance in that field.

The important point is that Chomsky's and Putnam's dispute isn't about
innateness per se. It's about whether language learning is modular. Putnam
would have spared the cognitive science community a fair amount of confusion if he'd named his target the *modularity* hypothesis (MH) rather than the
innateness hypothesis.

5.5   CHOMSKY ON LINGUISTIC UNIVERSALS

In this section, I aim to show that IH does, as Chomsky claims, explain LU.
Now expressions of the form "A explains B" are annoyingly ambiguous. On
the one hand, "A explains B" could mean that A is *the* one and only true
explanation of B; on the other hand, it could mean that A is *an* explanation
of B, in the sense that's compatible with there being alternative explanations
of B that are inconsistent with A. Each of these explanations in the second
sense would be adequate to serve as the correct explanation in the first sense,
although only one of them actually *is* the correct explanation. I will use "A
explains B" only in its second, non-exclusive sense. The discussion in this
section is designed to show that IH explains LU in this sense. It doesn't,
however, show that IH is the *correct* explanation. IH may even be false—and
while a false hypothesis may qualify as *an* adequate explanation of some
phenomena, it can never be *the* correct explanation of anything. Thus IH's
explaining LU doesn't rule out the possibility that there are other ways of
explaining LU.

Some of these alternative explanations of LU may be *better* than IH.
There's a considerable amount of controversy about what makes one explanation better than another. Also controversial is the principle of *inference to
the best explanation*, which asserts that the best explanation is the likeliest

to be true (other things being equal). There are some philosophers who will not grant the validity of this inferential rule (e.g., van Fraassen, 1980). But both Putnam and Chomsky accept it; so it's fair for either of them to use it in criticizing the other's views. According to this principle, an argument that A is the best explanation for B is a (defeasible) reason for inferring that A is the correct explanation, hence that A is true. Likewise, to argue that A is *not* the best explanation is to eliminate a potential reason for accepting it. Putnam's strategy is to show that there's a better explanation for LU than IH. If he's right, then he will have neutralized a potential reason for supposing that IH is true.

Now let's see how IH provides an explanation for LU. The built-in information postulated by IH can be thought of as a partial specification of the grammar of the language. In other words, part of the job of constructing the grammar is already done for us. But how can this be? Some of us—those of us who are being raised in Japan—are embarked on the project of learning Japanese, while others of us are learning German. If the built-in information is to be of help to the Japanese learner, it has to be a partial specification of Japanese; likewise, learners of German can only be helped by a portion of German. How can this be arranged? Supernatural agencies aside, there is no process that can selectively endow children with a partial knowledge of Japanese if they're *going* to grow up in Japan, and install a partial knowledge of German if they're *going* to be German. The only way to avoid an appeal to prescience is to assume that there are portions of Japanese that are identical to portions of German. Japanese and German children end up speaking different languages because they supplement this common innate endowment with different linguistic data.

But what about those of us who are learning Samoan? By the same reasoning, the innate endowment must simultaneously be a portion of Samoan as well as Japanese and German. The inevitable conclusion is that there are grammatical rules that apply to all the languages that human beings have ever learned as a first language.[12] Chomsky refers to the collection of all these rules as *universal grammar* (UG) (1980a, p. 28). Clearly, UG is a type of LU. Thus IH *predicts* that there are linguistic universals, which in turn means that the existence of linguistic universals is *confirmatory evidence* for IH. I use the term "predicts" in a loose (but not uncommonly encountered) sense. Strictly speaking, LU couldn't have been predicted by IH since it was antecedently known to be true. Strictly speaking, LU was *postdicted* by IH. But the conclusion is the same: LU is evidence for IH by virtue of being derivable from IH.

Putnam never called to question the Chomskyan thesis that LU is derivable from IH. He just argued that IH isn't the *best* explanation of LU. But the inferential bond between IH and LU has been challenged by Cowie (1999) in her analysis of the nativism-versus-empiricism issue. The main thesis of Cowie's book is that Chomskyan nativism consists of several partially independent hypotheses. The most grievous error on both sides of the dispute (according to Cowie) is the presumption that one must either accept or reject the entire nativist package. Among the several hypotheses that she claims are confounded in the nativist literature, she distinguishes the Universal Grammar hypothesis (U) from the Domain Specificity hypothesis (DS). DS is defined as follows:

> Learning a language requires that the learner's thoughts about language be constrained by principles specific to the linguistic domain (Cowie, 1999, p. 155).

And here is U:

> The constraints and principles specified in (DS) as being required for language learning are to be identified with the principles of the Universal Grammar (p. 157).

A small point on the nomenclature: Chomsky's UG is a set of rules, whereas Cowie's U is the hypothesis that these rules function as constraints on language learning.

Except in one respect, DS is the same proposition as the one that Putnam and I have called IH. The difference is that IH includes the specification that the constraining linguistic principles are innate, whereas DS leaves it open whether the constraining principles are innate or learned. For Cowie, innateness (I) is another partially independent nativist hypothesis that needs to be differentiated from its conceptual cousins. Here it is:

> The constraints and principles specified in (DS) as being required for language learning are in some manner innately coded (Matthews 2001, p. 217).[13]

So IH is the conjunction of DS and I.

According to Cowie, DS & I (a.k.a. IH) does *not* entail U:

> the case ... for (DS) and (I) falls short of establishing Chomsky's (U). That is, to argue that children must have *some* innate knowledge about natural language in order to learn one would seem to be one thing; and to make a case for a particular conception of what that knowledge is would seem to be another (1999, p. 174).

Cowie is not interested in criticizing or weakening the case for IH at this juncture. Her remarks are made in the service of her project of disassembling the nativist monolith. But, quite incidentally, her claim threatens to render Putnam's argument superfluous: you can't—and don't need to—argue that there's a better theoretical account of LU than IH if IH doesn't account for LU in the first place.

Is it possible for U to be false if it's granted that IH is true? The DS portion of IH stipulates that the learner's thoughts about language are constrained by principles specific to the linguistic domain, and the I portion of IH requires that these constraints be innate. Can there be constraints that meet these conditions but that don't conform to the principles of universal grammar? Well, we can imagine that the learner's thoughts about language are constrained by innate principles that are specific to French, but that are not applicable to any other natural language. In this case, it appears to be obvious that the linguistic constraints don't conform to the principles of UG. Yet there doesn't seem to be anything self-contradictory about the scenario of the French constraints. So it seems that Cowie's claim is vindicated: DS doesn't entail U. The French scenario has a peculiar property, however. The French constraints may or may not be adequate for success at learning French—but they're sure to be useless if the task is to learn English. Moreover, recall that DS stipulates that reliance on linguistic constraints is the *only* way that the learner can learn her first language. It seems to follow that if our linguistic constraints were the French constraints, we would be unable to learn English. But the task is to show that there is a body of linguistic information which is (1) not knowledge of UG, yet (2) sufficient for learning any and all natural languages. The next problem is to account for how we can come to learn English when we're endowed with the French constraints. I can think of only two ways that don't involve an appeal to occult forces.

First, English may be learned exclusively as a *second* language. For this possibility to be realized, English would have to be *constructed* by speakers who had already mastered a first language. But then English wouldn't be a natural language. Consequently, the fact that the French constraints don't apply to English doesn't make them less universal: UG isn't *supposed* to cover artificial languages. It's the same story with all the other non-French languages to which the French constraints don't apply. Given DS & I, the only way to learn a non-French language is as a second language—and then it doesn't have to be one of the languages that participate in UG. If we had French constraints, French would be the only natural language, and the French constraints would *be* the universal grammar. So this branch of the analysis doesn't establish the possibility that we learn language by relying on innate linguistic constraints, but that these constraints are not a part of the universal grammar.

The second way for there to be English speakers in a French-constrained world is for different human beings to possess different linguistic endowments:

some people have constraints that apply only to French, and the rest have constraints that apply to English. The second group may have constraints that apply *only* to English, or they may have the truly universal constraints that apply to all natural languages. In either case, the second group can learn to speak English as a first language but the first group can't. Members of the French-constraint contingent can, however, learn English as a second language from members of the second group. Looking at this imaginary society as a whole, we might allow that English is being learned as a first language even though there are linguistic constraints that don't apply to English. The transparent trick, however, is that those who can learn English as a first language and those who lack the English constraints are two distinct subpopulations. To foreshadow a theme to come, the two subpopulations are akin to two different language-using species, like human beings and Martians. Chomsky's theoretical views don't commit him to the thesis that the languages of every intelligent species in the universe obey the same constraints despite the fact that these species are the products of entirely different evolutionary histories. The rules of UG aren't intended to be literally universal. UG is a planetary grammar—and not even that if dolphins can talk. As for the scenario of the French constraints, the proper Chomskyan thing to say is that the two subpopulations have different UG's. One subpopulation's UG consists of sufficient information to reconstruct both French and English when that information is supplemented with the linguistic data that are obtained from close proximity to French and English speakers. The other subpopulation's UG consists, oxymoronically, only of the information that's sufficient for learning French.[14] So we haven't yet seen an example of linguistic information other than UG that's sufficient to enable us to learn every natural language.

A proponent of Cowie's thesis might retort that I've merely shown that having French-only constraints doesn't fit the bill—and that doesn't prove that there's no other package of non-UG information that *will* fit the bill. I agree. I presented the argument for the special case of French constraints in order to bring home the implausibility of Cowie's thesis. The general argument that DS & I entails U runs as follows. Assume DS & I. Then our innate knowledge entails sufficient information about English to enable us to figure out the grammar of English from the available linguistic data. The same can be said about every natural language. Thus our ability to learn any natural language is based on innate knowledge that applies to every one of these languages. That innate knowledge *may* consist of a direct description of grammatical rules that apply to all the languages. In that case, the innate knowledge is transparently knowledge of UG. Now it's true that the innate knowledge that enables us to learn languages need not logically be represented in the form of universal rules of grammar. Perhaps this is what Cowie had in mind when she claimed that DS and I doesn't entail U. But if our knowledge does in fact enable us to learn languages, it must *entail* those rules. Cowie defines U as the thesis that the principles of DS are "to be identified" with the rules of Universal Grammar. If the linguistic information in DS *entails* the rules of UG but isn't *identical* to the rules of UG, then U would indeed be false. But this is a mere technicality. Chomsky's monolithic nativism, as well as the claim that IH entails U, can be preserved by making an insignificant change to the definition of UG. We need only say that what gets into UG is any linguistic rule that's a consequence of our innate knowledge. If this definition is adopted, we may have to say that the rules of UG are nowhere represented in the mind of the learner. But so what? That was an interpretative option right from the start.

I conclude that the general argument to the effect that IH entails U stands unchallenged, and that therefore IH provides us with an adequate theoretical account of the existence of linguistic universals. But it isn't the only one.

5.6 PUTNAM ON LINGUISTIC UNIVERSALS

The rival explanation for the existence of linguistic universals is the *common origin* hypothesis (CO). According to CO, language was invented only once and then diverged into myriads of forms through a process of gradual historical change. On this account, all human languages have common features simply by virtue of their derivation from the same ancestor. There's some question whether CO does explain LU. It's obviously true that common descent results in common features over a span of several millennia. This is well illustrated by the pervasive similarities of the modern descendants of Latin. But to account for anything remotely like a *universal* similarity, we would have to go back to prehistoric ancestor-languages. According to Steven Pinker, the relation between common origin and shared features is eradicated on this time scale (Pinker, 1995, pp. 266–267). It's to be expected that the similarities between languages sharing a common ancestor will be attenuated with the passage of time. Thus it's also reasonable to suppose that there's a period of time beyond which the similarities due to common ancestry peter out completely. If we additionally have independent evidence that this period of time has been exceeded, then CO doesn't explain LU—and then this phase of Putnam's anti-Chomskyan operations doesn't get off the ground. It seems to me that making the relevant determination would be extremely difficult, especially in light of the fact there still are universal linguistic similarities.

But Putnam's argument that CO is a better explanation than IH fails whether or not it's conceded that CO explains LU. If CO isn't even *an* explanation for LU, then it trivially can't be a better explanation than IH. So we might as well grant Putnam the assumption that CO *is* an explanation of LU, and proceed to show that his reasons for supposing that CO is a *better* explanation than IH are defective.

I will discuss Putnam's argument in two parts. Here are both parts:

[I] Suppose that language-using human beings evolved *independently* in two or more places. Then, if Chomsky were *right,* there should be two or more *types* of human beings descended from the two or more original populations, and normal children of each type should fail to learn the languages spoken by the other types. Since we do not observe this . . . we have to conclude (if the I.H. is true) that language-using is an evolutionary 'leap' that occurred only *once.* But in that case, it is overwhelmingly likely that all human languages are descended from a single original language.

[II] . . . this hypothesis—a single origin for human language—is certainly *required* by the I.H., but much weaker than the I.H. But just this *consequence* of the I.H. is, in fact, enough to account for "linguistic universals"! (Putnam 1967/1980, pp. 246–247).

The first part of the argument purports to be a proof that IH, together with certain auxiliary assumptions, entails CO. The second part of the argument purports to show that the conclusion of the first part in turn entails that CO provides a better explanation for LU than IH does. For Putnam's purpose, which is to criticize IH, the analysis in the first part merely sets the stage for the main point to come. For *my* purpose, however, which is to trace the intricate web of ideas that surrounds this argument and to draw out their consequences for extraterrestrial communication, most of the interest lies in the first part. Following the expository strategy of leaving the best for last, I will discuss Putnam's issue first, and then turn to my own concerns. This amounts to dealing with the second part of the argument first. I begin by assuming that Putnam has achieved his aim in the first part—he's shown that IH entails CO. The status of this assumption will be examined in the next section. The topic for this section is whether Putnam succeeds in achieving his aim in the second part, which is to show that IH's entailing CO in turn entails that CO is a better theory than IH.

Here is Chomsky's response to Putnam's argument:

Let me then turn to Putnam's . . . argument, that even if there were surprising linguistic universals, they could be accounted for on a simpler hypothesis than that of an innate universal grammar, namely, the hypothesis of common origin of languages. This proposal misrepresents the problem at issue . . . [T]he empirical problem we face is to devise a hypothesis about initial structure rich enough to account for the fact that a specific grammar is acquired, under given conditions of access to data. To this problem, the matter of common origin of language is quite irrelevant. The grammar has to be discovered by the child on the basis of data available to him, through the use of the innate capacities with which he is endowed. . . . [The child] knows nothing about the origin of language and could not make use of such information if he had it (Chomsky, 1969/1981, p. 302).

Thirty-two years later, the Chomskyan critique of Putnam's argument is essentially the same:

even if linguistic universals could be explained by common descent, the explanation would be beside the point in the present context. This is because the Poverty of the Stimulus Argument isn't particularly concerned with linguistic universals. It's about the situation that every child faces as she comes to the task of acquiring language. As Chomsky (1967) notes, common descent can't help the child at all; linguists may know about the descent of a target language, but children most certainly do not (Laurence & Margolis, 2001, p. 248).

The recent critique recapitulates two infelicities in the original. The critics start off on the right footing when they note that CO is "quite irrelevant" to the problem of language acquisition (Chomsky), or that "common descent can't help the child" to acquire language (Laurence & Margolis). The point is obvious: the history of a language has nothing to do with the problems the child has to face in trying to learn it. Having made this valid point, the critics' next sentence is presented as though it were an elaboration of the same idea. But in fact they both make a peculiar switch. Chomsky tells us that "the child knows nothing about the origin of language and could not make use of such information if he had it," and Laurence and Margolis write that "linguists may know about the descent of a target language, but children most certainly do not." In both cases, the critics have switched from making the relevant point that common descent can't help language acquisition to the *ir*relevant point that *knowledge* of common descent can't help language acquisition. The second point is undoubtedly true, but its truth or falsehood has no bearing on the status of CO. More generally, there's no reason to expect any systematic relationship between the consequences of a theory and the consequences of knowing the theory. Suppose that someone, in the throes of a monumental confusion, suggests that quantum mechanics explains language acquisition. The way to unconfuse him would be to persuade him that the principles of quantum mechanics don't make conceptual contact with the problems of language learning. That is to say, we would talk to him about the theory. It would *not* be relevant to point out that the language learners don't know quantum mechanics, or that a knowledge of quantum mechanics wouldn't help them in their task. Quantum mechanics is one thing; knowledge of quantum mechanics is another. The same can be said about CO and knowledge of CO.

Still, I've conceded that the critics also make the relevant point that CO itself—not knowledge of CO—doesn't explain language acquisition. Doesn't this mean that we obtain an adequate refutation of Putnam's position if we simply cross out the unfortunate remarks about knowledge of CO? Well, we get one step of a refutation, but there's a lot more that needs to be said. Apparently, the critics have lost track of what Putnam is trying to do. He does *not* try to claim that CO explains language acquisition, much less that it's a better explanation of this phenomenon than IH. His claim is that CO is a better explanation *of LU*. It's true that his ultimate objective is to discredit IH—but here he pursues this goal with a certain amount of indirection. IH is at least a viable candidate-explanation of LU. If it could be established that IH is the correct explanation for LU, we would also establish that IH is true. Even if it could only be established that IH is the best of the available candidate-explanations of LU, that would be a powerful reason for accepting IH. If Putnam succeeds in showing that CO is a *better* explanation for LU, he effectively deprives IH of these epistemic benefits.

Both Chomsky and Laurence and Margolis mention the fact that CO doesn't explain language acquisition without deriving any further consequences from this fact or conjoining it with any other considerations. This strongly suggests that the bare fact about CO is already considered to be an adequate refutation of Putnam: the latter's concern with CO is irrelevant to the status of IH because CO doesn't speak to the issue of language acquisition, which is what IH is all about. But theories may be criticized and even overturned on the basis of issues far removed from the concerns that led to their formulation in the first place. For example, the discovery of a contradiction between a theory and an established fact would precipitate a crisis for the theory, even if the fact were about matters that are very distant from the theory's usual concerns. So while it's true that IH was devised primarily to explain language acquisition, and it's true that CO makes no contact at all with the topic of language acquisition, it's nevertheless possible for a fact about CO to constitute an effective critique of IH. To be sure, Putnam's argument doesn't show that IH is false; but if it succeeds, it eliminates a potential reason for adopting IH.

Putnam's argument is straightforward. He takes it for granted that both CO and IH are viable explanations of LU. He's also argued that IH entails CO, and we've agreed to accept this claim for the time being. It follows that

CO is a *weaker* hypothesis than IH, in the sense that IH says more about the world than CO does: by virtue of entailing CO, IH asserts everything that CO asserts, and it says more besides. Putnam then invokes a general methodological principle used routinely by theoreticians in all fields of science: given a choice between two theoretical explanations of the same phenomena, we should, *ceteris paribus*, prefer the weaker one. This is not a principle to which Chomsky would object. Yet it seems to lead directly to a preference for CO over IH. So Putnam wins this round.

Or does he? There is a counterargument available to Chomsky of which his observation that CO doesn't explain language acquisition is only a part. There's a *ceteris paribus* clause in the principle that Putnam invokes: we should prefer the weaker of two theories *other things being equal*. Now one way in which theories may *not* be relevantly equal is in their *explanatory scope*. If theory T1 explains more than T2, that's a theoretical virtue that may counterbalance the deficit of being stronger than T2. Like the principle invoked by Putnam, this one enjoys frequent and uncontroversial employment in all the sciences. In fact, it's invoked every time a theory is put forth that has implications that go beyond the data. If relative strength were the only measure of theoretical virtue, we would never adopt any theory other than a catalogue of the data themselves. To say anything more is to make one's claims stronger. Nevertheless, we often do formulate theories that go beyond the data, the rationale being that what we lose in parsimony is offset by a gain in explanatory scope.

In the case at hand, it's true that CO is a weaker theory of LU than IH is. But it also has a narrower explanatory scope. IH explains both LU and language acquisition, whereas CO explains only LU—*CO doesn't explain language acquisition*. This last statement is the totality of Chomsky's and of Laurence and Margolis' rebuttal. But it accomplishes nothing in isolation. It's only when it's embedded in the broader argument that it effectively blocks the force of Putnam's argument. Whether the greater explanatory scope of IH is of sufficient magnitude to offset the greater parsimony of CO is a judgment call on which reasonable people might disagree. There are no known algorithms for making these decisions. But at least it can be said that Putnam hasn't made his case.

A quite different criticism of Putnam's argument is offered by Cowie (1999). I cited this author as a contemporary representative of the anti-Chomskyan persuasion. However, she isn't impressed by what Putnam has to say about linguistic universals:

Putnam also speculates that linguistic universals might be due to all languages' having evolved from a single ancestral language. There is, however, no uncontested linguistic or historical evidence for the existence of a common Ur-language. Rather, the consensus seems to be that language has probably evolved separately in a number of different places (Cowie, 1999, p. 180).

On this account, Putnam's critique of Chomsky's claim that IH is the correct explanation of LU amounts to noting that the explanation of LU might just as well be CO. Cowie's critique of Putnam's critique is that there's no independent evidence that would warrant our accepting that CO is true. This is a drastic misrepresentation of Putnam's argument. The truth or falsehood of CO doesn't figure in his argument. He claims that the acceptance of IH forces Chomsky also to accept CO—and then it turns out that CO is a better explanation of LU. There is no place in this critique where Putnam has to invoke the truth of CO. Putnam's view is that IH fails whether or not CO is true. If CO is true, then it's a better explanation of LU than IH is—this is the argument that I criticized above. What if CO is false? According to Cowie, Putnam's feeble attack on IH would thereby be completely deflected. But in fact, the falsehood of CO makes the case against IH even stronger. If Putnam's claim that IH entails CO is allowed to stand, then a demonstration that CO is false also shows by *modus tollens* that IH is false!

5.7   PUTNAM'S LEMMA

I turn now to the first part of Putnam's argument, in which he claims to establish enough to infer that CO is a weaker theory than IH. The analysis begins:

Suppose that language-using human beings evolved *independently* in two or more places. Then, if Chomsky were *right*, there should be two or more *types* of human beings descended from the two or more original populations, and normal children of each type should fail to learn the languages spoken by the other types.

Let E2 be the hypothesis that language evolved independently in two or more places, and let AL be the hypothesis that all normal children can learn any human language. The assumption that Chomsky is right is, of course, IH. The whole passage can be represented as:

$$(\text{IH \& E2}) \rightarrow \sim\!\text{AL}. \tag{1}$$

Here's the case for (1) in greater detail. (Big caveat: this reformulation intentionally includes the errors and oversights in the original, so please hold all criticisms until the end.) Assume that IH is true. Then nobody can learn a language without relying on innate grammatical information. Now assume E2—that language-using human beings evolved twice. Let A and B be the two groups of humans in which these evolutionary developments took place. Each evolutionary event will implant its own distinctive package of innate linguistic information. Let I(A) and I(B) represent the packages implanted in groups A and B, respectively. Furthermore, let L(A) be the set of all languages whose grammar can be derived by supplementing the linguistic data with I(A), and let L(B) stand in the same relation to I(B). Children who are members of A will be able to learn the languages of L(A), for they possess the requisite innate information I(A). However, lacking the requisite innate information L(B), they *won't* be able to learn the languages of L(B). Symmetrical considerations show that members of B won't be able to learn any of the languages in L(A). Thus A and B comprise two groups of children whose linguistic capacities are entirely disjoint: the languages accessible to each group are inaccessible to the other. That is to say, ~AL. This completes the proof of (1).

Putnam continues:

> Since we do not observe this . . . we have to conclude (if the I.H. is true) that language-using is an evolutionary "leap" that occurred only *once*.

This passage asserts the following:

$$(AL \ \& \ IH) \rightarrow {\sim}E2. \text{[15]} \tag{2}$$

Proposition (2) does indeed follow from proposition (1). The passage asserts further that AL is true, and correctly infers from this and from (2) that IH → ~E2—that if the innateness hypothesis is true, then language evolved only once.

The finale:

> But in that case, it is overwhelmingly likely that all human languages are descended from a single original language.

This is, of course CO. It's asserted to be an overwhelmingly likely consequence of ~E2, which is in turn a consequence of IH (given that AL is true). But if IH → ~E2 and ~E2 → CO, it follows that

$$IH \rightarrow CO. \tag{3}$$

This is the result that's the basis of Putnam's claim in the second half of the argument that CO is a weaker theory than IH. Let's call it *Putnam's lemma*.

In the previous section, I objected to Putnam's inference from (3) to the conclusion that CO is the better of the two theories. I accepted the inference from (3) to the proposition that CO is the *weaker* of the two theories, but denied Putnam's claim that CO's being the weaker theory justified a preference for it. In this section, I want to mount another counterargument against Putnam. I claim that he fails to establish that CO is the weaker theory. It's undoubtedly true that COs being the weaker theory is a consequence of (3). But Putnam's argument doesn't establish (3), at least not in a sense that licenses the inference that CO is weaker. By his own account, he shows that IH entails CO *given that any human child can learn any human language*—that is, given AL. Formula (3) is the conclusion of an argument that employs AL as a premise. Now, supposing that his argument is sound, it's perfectly permissible for Putnam to assert its conclusion, (IH $\rightarrow$ CO), detached from the premise AL. But you can't ignore the premises when you're making judgments of logical strength. Here's an illustration of what can happen if you do. Given the premise that all non-Belgian emeralds are green, it follows that "All Belgian emeralds are green" entails "All European emeralds are green." By Putnam's reasoning, we would have to conclude that "All European emeralds are green" is weaker than "All Belgian emeralds are green," which is obviously wrong. What's justified in this illustrative case is the conclusion that "All European emeralds are green" is weaker than "All Belgian emeralds are green," *together with the premise that all non-Belgian emeralds are green*—in other words, that "All European emeralds are green" is weaker than "All emeralds are green." By the same token, tucking the premise into the antecedent of (3) yields

$$(\text{IH \& AL}) \rightarrow \text{CO.} \tag{4}$$

This formula licences the inference that CO is weaker than (IH & AL), but it doesn't follow that CO is weaker than IH itself. If "Putnam's lemma" is intended to name the relationship between IH and CO, then (4) is a corrected version of the erroneous version given in (3).

To summarize what's been accomplished so far: Putnam's argument that CO is a better explanation than IH for LU is vindicated from Chomsky's charge, but stands accused of two other charges. It's vindicated from the criticism that CO doesn't explain language learning: it doesn't have to explain language learning itself in order to stand as a rebuke to a theory of language learning. The two new charges are (1) that COs being the weaker theory doesn't in this case warrant the inference that it's a better theory (see the previous section), and (2) that Putnam fails to show that CO is the weaker theory in the first place (see the previous paragraph).

5.8 THE EVOLUTION HYPOTHESIS

The derivation of CO from IH also employs the premise ~E2 that language evolved only once. Putnam claims that ~E2 follows from IH & AL (see formula (2)). If he is right, then the latest version of Putnam's lemma (formula (4)) doesn't need amendation—its present antecedent IH and AL already contains the needed ~E2. But Putnam's discussion of (2), as well as my own plodding version of that discussion, glosses over a significant assumption. Both versions ask us to assume E2, that language evolved twice, and they infer from that assumption together with IH that ~AL—that the descendants of the two lineages will be unable to learn each other's languages. This inference relies on the tacit assumption that each evolutionary leap implants a different package of innate linguistic information. In fact, the argument seems to require that the two sets of innate constraints I(A) and I(B) be mutually inconsistent, in the sense that any language that satisfies I(A) must fail to satisfy I(B) and vice versa. For suppose they aren't mutually inconsistent. Then there are possible languages that belong to both L(A) and L(B). If members of B should happen to speak such a language, it might be possible, contra Putnam, for members of A to learn the language of the B's after all. We would then have failed to establish ~AL as a consequence of ~E2. Let's call the assumption presupposed by Putnam—that two evolutionary leaps into language would implant mutually incompatible packages of innate linguistic information—the *evolution hypothesis (EH)*.

It's at least *prima facie* plausible that EH is true—for there are surely many sets of grammatical rules that serve the purpose of language equally well. Over a broad range of possibilities, the choice of grammatical rules is rather like the choice between driving on the right-hand side of the road or driving on the left. What matters is only that we do the same thing. There being little or no adaptive advantage to one set of rules over another, there's no reason to expect that evolutionary processes are going to select the same rules twice. On the other hand, Chomsky reminds us that we don't really know enough to have a firm opinion of these matters:

> In studying the evolution of mind, we cannot guess to what extent there are physically possible alternatives to, say, transformational generative grammar, for an organism meeting certain other physical conditions characteristic of humans. Conceivably, there are none—or very few . . . (1972, pp. 97–98).

If, as Chomsky speculates, there are only a few nomologically possible sets of innate constraints, then it may not be improbable for two independent selections to overlap. To be on the safe side, we should include EH as one of the conditions that need to be met for a successful derivation of CO from IH. Putnam's lemma becomes:

$$(IH \ \& \ AL \ \& \ EH) \rightarrow CO. \qquad (4)$$

5.9 IH*

But expanding the antecedent of Putnam's lemma from IH to (IH & AL & EH) still doesn't make it strong enough to derive CO. Recall the structure of the argument for IH: language acquisition can't be accounted for by rational inference from the available linguistic data, but it can be accounted for by IH. Let's grant that this is so. In fact, let's grant that the innateness explanation is true: children learn their language by relying on innate linguistic information. To say this is not yet to say that there isn't a *third* type of explanation that also describes a process whereby human children *might* learn a language. The fact that all of us have always learned our language by relying on innate linguistic information doesn't yet entail that we couldn't also have learned our language in some other way. But if there *is* some other way to learn a language, then the argument for Putnam's lemma is inconclusive. The fact that members of the A-group lack the requisite innate information I(B) may entail that they can't learn the languages of L(B) *by the procedures specified in Chomskyan explanations.* But if there is another way to learn languages, it may yet be possible for A-children to learn to speak B-ish. It may be that we learn our languages by relying on the procedures specified in Chomskyan explanations whenever we can, but that when these procedures fail us, we have an alternative procedure to fall back on. It's just that the procedures specified in Chomskyan explanations have never failed us yet.

In any case, for Putnam's lemma to go through, we need to assume more than that the innateness hypothesis is true. The oversight can be clearly seen at the very beginning of my plodding version of Putnam's argument. I begin by assuming that IH is true, and I interpret this to mean that no one can learn a language without relying on innate grammatical information. But that isn't what IH says. What IH says is that children *do* learn language by relying on innate grammatical information. To get the stronger claim, you have to assume more than IH. You need to assume that the innateness hypothesis describes the only way in which human children *can* learn their language. Let's represent this assumption by IH*. Then Putnam's lemma is the following proposition:

$$(IH \ \& \ AL \ \& \ EH) \rightarrow CO. \qquad (5)$$

The distinction between the weak and the strong arguments for IH figures here. The weak argument is based on the premise that the linguistic data that are in fact available to children are insufficient for a rational reconstruction of the language. The strong argument maintains that any collection of linguistic data that *could* be made available to children is insufficient for a rational reconstruction. Either argument can be used to support IH: whether the insufficiency of data is *de facto* or *de jure*, it created a problem that IH resolves. The two arguments have very different relationships to IH*, however. If you endorse only the weak argument, you're allowing that there might be a package of linguistic data which would enable children to learn the language without the assistance of innate knowledge. If you endorse the strong argument, you rule out this possibility. Now IH* is the thesis that there is no way of learning a first language without innate knowledge. Thus, to accept the strong argument is to improve IH*'s epistemic standing by eliminating a potential counter instance. But the weak argument has nothing to say about the existence of such a counter instance. In sum, the strong argument strengthens the case for IH*, whereas the weak argument doesn't.

5.10   HISTORICAL PREMISES

We need one more addition to our ever-inflating antecedent. The current antecedent, IH* + AL + EH, is strong enough to yield the conclusion ~E2: if reliance on innate constraints is the only way to learn a language (IH*), and if two evolutionary leaps into language produce two mutually incompatible sets of constraints (EH), and if all children can learn any language (AL), then language evolved only once (~E2). But what about the conclusion that all languages have a common origin (CO)? Putnam writes that ~E2 renders CO "overwhelmingly likely." I don't see how this can be anything more than a wild guess. It's easy to construct plausible scenarios wherein ~E2 is true and CO is false. For example, it might have happened that the capacity to use language evolved only once (~E2), but that a considerable period of time elapsed before this capacity started to be utilized. Meanwhile, the language-ready descendants of the first language-ready human could have dispersed and formed isolated groups, and two or more of those groups could then have constructed languages independently of one another. It might be objected that it's implausible that an innate capacity for speaking should remain unused for a substantial period. To this there are two replies. First, there are reasons to suppose that the capacity for language *couldn't* get utilized until a substantial period of time had elapsed. If Wittgenstein's *private language argument* is right, there cannot be a language that's spoken by only a single person—language can only be possessed by a community (Wittgenstein, 1953). Then the first language-ready human would have had to remain mute until her offspring comprised a community. But that would have provided enough time for a dispersal to take place that would produce several non-communicating communities, in each of which a different language might develop *ab ovo*.

Second, even if a delay between evolutionary endowment and utilization proves to be impossible or implausible, there are ways in which CO might fail that don't require a delay. One quick example: a non-language-ready tribe raids the tribe containing the world's first language-ready human, kidnaps her newborn language-ready child, and raises the child as its own. The mother and child can them become loci for the development of two independent Ur-languages. I don't have a firm opinion of how likely these just-so stories may be; but Putnam gives us no reason to suppose, as he does, that they're overwhelmingly *un*likely. In any case, we can't go wrong by adding Putnam's opinion as a premise to his derivation of CO from IH. Let *E1* be the hypothesis that *all the existing languages rendered learnable by a single evolutionary leap have a common origin*. Then Putnam's lemma becomes:

$$(\text{IH}^* \ \& \ \text{AL} \ \& \ \text{EH} \ \& \ \text{E1}) \ \rightarrow \ \text{CO}. \qquad (6)$$

Formula (6) is, I think, correct as it stands: if deploying one's innate linguistic knowledge is the only way to learn a language (IH*), and any human child can learn any human language (AL), and two evolutionary leaps into language result in incompatible packages of innate knowledge (EH), and all the human languages rendered learnable by a single evolutionary leap have a common origin (E1), then all human languages have a common origin (CO). Formula (6) is the lemma that Putnam has to carry into the second part of the argument against Chomsky. It's the long antecedent of (6), (IH* & AL & EH & EI), that's logically stronger than CO. IH alone is a tiny logical fraction of this proposition, and there's no reason to believe that the strength of this fraction exceeds the strength of CO.

5.11   THE CETI CONSEQUENCE

Note that neither Chomsky's counterargument to Putnam's critique nor my expanded version of the counterargument involves the rejection of Putnam's lemma. Chomsky merely repudiates the *further* consequence that the common origin hypothesis is superior to the innateness hypothesis. But it's precisely this unrefuted portion of Putnam's argument that bears on the issue of extraterrestrial communication. An almost complete argument that we will never be able to decode an extraterrestrial language (if Chomsky's theory is right) is obtained by substituting the more generic "beings" for the several occurrences of the words "human beings" and "children" in the original statement of Putnam's lemma:

> Suppose that language-using beings evolved *independently* in two or more places. Then, if Chomsky were *right*, there should be two or more *types* of beings descended from the two or more original populations, and normal beings of each type should fail to learn the languages spoken by the other types.

Clearly, this argument succeeds or fails together with the original argument for Putnam's lemma. To obtain the terms of the new argument, simply cross out all occurrences of the word "human." The passage adapted from Putnam then says: assume ~CO; then, if IH*, then ~AL. This comes to the same thing as proving that ~CO $\rightarrow$ (~IH* v ~AL), which is the same thing as (IH* & AL) $\rightarrow$ CO. When the terms had their old, anthropocentric meanings, this formula was called Putnam's lemma; with the new meanings, it's to be called the *CETI consequence*. Because of the differences between my theoretical agenda and Putnam's, the contrapositive of Putnam's lemma is a more perspicuous statement of the CETI consequence than is the lemma in its original form: Putnam wants to derive CO from IH & AL, whereas I want to derive ~AL from IH* & ~CO. For present purposes, the best statement of the CETI consequence is

$$(\sim\!CO \ \& \ IH^*) \rightarrow \sim\!AL. \tag{7}$$

This formula may be read as follows: if possessing innate linguistic knowledge is the only way to learn a first language, then two or more groups of beings that developed language independently will be unable to learn each other's language. More telegraphically: if Chomsky is right, then CETI is doomed to failure.

It needs to be emphasized that the portion of Putnam's argument which is at stake—the portion dubbed "Putnam's lemma"—by itself poses no serious threat to Chomsky's theory. If Putnam's lemma should prove to be sound, Chomsky would only have to abandon his aversion to the common origin hypothesis and to admit that its truth is at least as likely as the truth of the innateness hypothesis. All his other opinions could remain unchanged. The issue doesn't loom large on Chomsky's theoretical agenda. But when it comes to the agenda of CETI, the point becomes crucial. If Putnam's lemma is accepted, then it seems but a small step—a mere deletion of the superfluous adjective "human"—to arrive at the conclusion that a highly influential scientific theory—Chomsky's—entails that CETI must fail.

Naturally, the CETI consequence is subject to the same additional provisos, EH and E1, as Putnam's lemma. Its fullest and most perspicuous form is a contrapositive of (6):

$$(\sim\!CO \ \& \ IH^* \ \& \ EH \ \& \ E1) \rightarrow \ \sim\!AL. \tag{8}$$

EH is the hypothesis that two different evolutions of language would implant two incompatible packages of innate linguistic information. E1 rules out scenarios wherein language is invented twice by groups that have the same evolutionary history. If condition E1 is not satisfied, we may very well be able to converse with extraterrestrials. Here is one series of circumstances in which E1 fails to be satisfied. Extraterrestrials visit our planet soon after the modern *Homo sapiens* has evolved, and they abscond with a shipload of newborn human infants. When they're old enough to fend for themselves, the humans are released on an uninhabited Earth-like planet and left to their own devices. Why do the extraterrestrials do this? Who knows?—they're *extraterrestrials*. Anyway, in due time, the members of this human colony and their descendants develop a language and discover radio. Their transmissions are picked up by SETI researchers on Earth, who triumphantly—and quite accurately—declare their project to be a success: they've found extraterrestrial intelligence. But, the CETI consequence notwithstanding, these extraterrestrials speak a language that we can hope to understand. This follows from the facts that (1) these extraterrestrials have an innate endowment that enables them to learn their language, and (2) we have the same innate endowment as they do.

5.12   THE WEAK AND STRONG ARGUMENTS REVISITED

Recall the distinction between the weak and the strong versions of the argument for IH. The weak argument is that children *don't* receive enough linguistic data to reconstruct the grammar of the language by a process of rational inference. The strong argument is that children *can't* be given enough linguistic data to accomplish this feat. The weak argument provides just as much support for the existence of innate linguistic constraints as does the strong argument. But Putnam's lemma and the CETI consequence require the strong argument. Let's call a language that doesn't have the same evolutionary origin as one's native language an "alternative language." The weak Chomskyan argument licenses at most the following weakened version of Putnam's lemma—that children won't learn an alternative language *if they're given the kind of linguistic education that children typically receive*—that is, if they're given the usual amount and variety of linguistic data—the amount that we know to be insufficient for a rational reconstruction of the language. There's nothing in the weak argument that rules out the possibility that an extraordinarily long, strenuous, and systematic educational program might not enable the child to learn an alternative language for which it lacks appropriate innate constraints.

But if this is what the argument comes to, then CETI has nothing to fear from Chomskyan linguistics. The argument concedes that it's possible to learn languages that fail to satisfy our innate constraints. Maybe we just need a greater quantity and better quality of linguistic data than children are wont to obtain. Perhaps it would take a century of carefully designed instruction. We may be sure that our children didn't learn English in this manner. But there's no reason why we shouldn't be able to learn the language of extraterrestrials by the slow route. This conclusion need not discomfit the friends of CETI. Nobody ever said that CETI had to be easy. Of course, if learning an extraterrestrial language is *absurdly* difficult—on the order of, say, building a four-lane highway from the Earth to the moon—then its mere nomological possibility may not be enough of a reason to proceed.

5.13   CHOMSKY ON THE CETI CONSEQUENCE

There's a bit of indirect evidence concerning Chomsky's own view of CETI. It comes to us through Putnam, who tells us that he's discussed these issues with Chomsky many times (Putnam, 1967/1980, p. 240). The evidence suggests that Chomsky disagrees with my claim that his theoretical commitments entail the impossibility of CETI. Putnam:

> If intelligent non-terrestrial life—say, Martians—exists, and if the "Martians" speak a language whose grammar does not conform to [the universal grammar of human languages], them I have heard Chomsky maintain, humans (except possibly for a few geniuses or linguistic experts) would be unable to learn Martian; a human child brought up by Martians would be unable to learn language; and Martians would, conversely, experience similar difficulties with human tongues. (Putnam, 1967/1980, p. 241).

What is significant about this quote is the exception of a few geniuses. At first glance, this qualification seems quite unwarranted. We've seen that Chomsky explicitly, albeit infrequently, endorses the strong argument for innateness. But the conclusion of this argument is that the inability to learn alternative languages is irremediable. Having an IQ of 200 or 2000 won't enable you to reason your way from any collection of linguistic data to the correct grammar without the assistance of innate constraints. The following passage by Chomsky himself tells us what he has in mind:

> . . . a rich set of principles of universal grammar permits us to attain our extensive knowledge of language on limited evidence, and by the same token, these principles exclude languages that violate the principles as inaccessible to the language faculty (some might be learned, with effort, application, and explicit formulation and testing of hypotheses, by means of other faculties of mind) (1980a, p. 251).

Strictly speaking, this passage makes the explicit claim that we *can* learn alternative languages. Thus it's a direct repudiation of the strong version of Putnam's lemma. Its effect on the argument is minimal however, requiring only a minor adjustment in the formulation of the lemma. What Chomsky has in mind here is that even if a language falls outside of the constraints of the *language* module, it may turn out to be possible to learn it by co-opting a *different* cognitive system that functions in accordance with different constraints. He clarifies the point elsewhere:

> If, indeed, the mind is modular in character, a system of distinct though interacting systems, then language-like systems might be acquired through the exercise of other faculties of mind, though we should expect to find empirical differences in the manner of acquisition and use in this case (1980a, p. 28).

Thus the strong Putnam lemma—the thesis that children can't learn alternative languages—is strictly false on account of Putnam's having neglected this possibility. A child might after all learn an alternative language by means of another faculty. But this doesn't affect much of anything. For one thing, Putnam can still make the argument that the innateness hypothesis entails the common origin hypothesis. The argument just has to be amended a bit. The bare fact that all children can learn any human language doesn't yet do the job. But Putnam can import the stronger, but still obviously true, empirical premise that all children can learn any human language with more or less equal facility. For if the language module had developed twice, and a child was learning an alternative language by means of an imported module, the learning would require, as Chomsky says, extraordinary effort and application.

It's also clear that the proviso that we may be able to learn alternative languages by other modules doesn't jeopardize the CETI consequence. Grant first that a given module can only learn the languages that fit its constraints. Grant also that if a module evolves twice, it's going to have different constraints—that is, EH. Putnam's minor error is in supposing that if children have different *language* modules, then they won't be able to learn each other's language. This overlooks the possibility of learning the alternative language by means of an alternative module. But when we're dealing with extraterrestrials, *all* the modules are going to be different. So the possibility of alternative module learning doesn't change anything. The CETI consequence still follows.

To recapitulate: the strong Putnam lemma is false. Chomsky's theory does allow for the possibility that children can learn alternative languages. But for our purpose, it's false on a mere technicality. An amended form still yields the CETI consequence. The amended form is that if *minds*, considered as total aggregates of cognitive modules, evolved twice, then they wouldn't be able to learn each other's languages.

5.14 THE PURSUITWORTHINESS OF CETI RECONSIDERED

In sum, there's bad news and good news for CETI advocates. The bad news is that a canonical scientific theory makes the prediction that their goal can never be attained. The good news is that a canonical scientific theory predicts that their goal is unattainable. It's bad news for CETI on account of the fact that it provides rational grounds for scepticism about the prospects of CETI success. On the other hand, it's good news for CETI on account of the fact that it brings CETI fully into the realm of ongoing theoretical discourse. I had argued in section 2.2 that there's a technical, objective sense in which CETI and SETI are scientifically *boring*. What I had in mind was that the answers to the questions raised in the field of extraterrestrial studies would shed little or no light on any theoretical issues. To be sure, success at SETI or CETI would be *intrinsically* interesting, like news of the carryings-on of celebrities. But in both cases, the information obtained would neither confirm nor disconfirm any current theoretical hypothesis of physics, or chemistry, or biology, and so on. It seems that I have to retract this opinion. A CETI success would be strong evidence against what has become the orthodox theory of language acquisition. Even if we never encounter any language-using extraterrestrials, it illuminates the theory to know that this connection to CETI exists.

Notes

1. My historical information is drawn entirely from Beck (1985), Crowe (1986), and Dick (1982). These monographs should be consulted for references to original sources.
2. William Whewell belonged successively to both camps. See Crowe (1986).
3. A more elaborate version of the foregoing argument can be found in Kukla (1991).
4. For a history of the Drake equation, see Dick (1996).
5. Dick's historical summary of SETI project funding makes no mention of any grants for the investigation of psychosocial factors. See Dick (1996, pp. 494–501).
6. Recent evidence for the existence of extrasolar planets arguably warrants the conclusion that f_p, the fraction of stars bearing planets, is non-negligibly greater than zero. This is an essential step in making a case for the plausibility of ETI. But it does not, by itself, mitigate the critique of the current status of that case. The critique hits its target so long as there fails to be a systematic basis for estimating even a single Drake factor.
7. An excellent account of vague probabilities is given by Bas van Fraassen (1989).
8. In fact, the truth of LU is regarded as uncontroversial by the vast majority of linguists. See Maratsos (1989) for one of the few dissenting views.
9. I ignore here the problem of how the language acquisition device discriminates between speech acts and non-linguistic vocalizations such as peals of laughter or cries of pain.
10. I have distinguished the varieties of constraint styles more carefully in Kukla (1992).
11. See, for example, Chomsky (1962, p. 549).
12. Second-language learning is different because the learner has another source of information about the language that she's trying to learn. In at least some cases, she can be directly *told,* in her first language, what the rules of the second language are.

13. This is a reviewer's suggested substitute for Cowie's original formulation of I. In her response to the review, she indicates her full acceptance of the amended version (Cowie, 2001, p. 239).

14. If the information in UG applies only to French among the present natural languages, then the possessor of this UG won't be able to learn any of these present languages except French. But it's possible that another learnable language will develop in the future. This could happen if the new language obeyed the French constraints.

15. Strictly speaking, ~E2 includes the case where the evolutionary leap to language didn't occur at all. I ignore this innocuous complication in order to simplify the presentation.

References

Anderson, P. (1963). *Is there life on other worlds?* New York: Crowell-Collier.

Baird, J. C. (1987). *The inner limits of outer space.* Hanover: University Press.

Barrow, J. D., & Tipler, F. J. (1988). *The anthropic cosmological principle.* Oxford: Oxford University Press.

Beck, L. W. (1985). Extraterrestrial intelligent life. In E. Regis Jr. (ed.), *Extraterrestrials: Science and alien intelligence.* Cambridge: Cambridge University Press, pp. 3–18.

Cameron, A. G. W. (ed.). (1963). *Interstellar communication: A collection of reprints and original contributions* (New York: W. A. Benjamin).

Chomsky, N. (1962). Explanatory models in linguistics. In E. Nagel, P. Suppes, & A. Tarski (eds.), *Logic, methodology and philosophy of science* (pp. 528–550). Stanford, CA: Stanford University Press.

Chomsky, N. (1967). Recent contributions to the theory of innate ideas. In R. S. Cohen & M. W. Wartofski (eds.), *Boston studies in the philosophy of science, Vol. III.* Dordrecht, Holland: D. Reidel.

Chomsky, N. (1969/1981). Reply to Putnam. In N. Block (ed.), *Readings in philosophy of psychology, Vol. 2* (pp. 300–304). Cambridge, MA: Harvard University Press.

Chomsky, N. (1972). *Mind and language* (2nd ed.). New York: Harcourt Brace Jovanovich.

Chomsky, N. (1980a). *Rules and representations.* New York: Columbia University Press.

Chomsky, N. (1980b). Rules and representations. *Behavioral and Brain Sciences, 3,* 1–15.

Chomsky, N. (1986). *Knowledge of language: Its nature, origin, and use.* New York: Praeger.

Chomsky, N., & Fodor, J. (1980). The inductivist fallacy. In M. Piattelli-Palmarini (ed.), *Language and learning: The debate between Jean Piaget and Noam Chomsky* (pp. 255–275). Cambridge, MA: Harvard University Press.

Chow, S., & Teicher, H. (1978). *Probability theory: Interdependence, interchangeability, martingales.* New York: Springer-Verlag.

98   *References*

Clarke, A. C. (1973) *Report on planet three*. New York: Signet.

Cowie, F. (1999). *What's within? Nativism reconsidered*. New York: Oxford University Press.

Cowie, F. (2001). Cussing in church: In defense of *What's 'within*. *Mind and Language, 16*, 231–245.

Crowe, M. J. (1986). *The extraterrestrial life debate 1750–1900: The idea of a plurality of worlds from Kant to Lowell*. Cambridge: Cambridge University Press.

Davidson, D. (1974). On the very idea of a conceptual scheme, *Proceedings and Addresses of the American Philosophical Association 67*, 5–20.

Dick, S. J. (1982). *Plurality of worlds: The origins of the extraterrestrial life debate from Democritus to Kant*. Cambridge: Cambridge University Press.

Dick, S. J. (1996). *The biological universe: The twentieth-century extraterrestrial life debate and the limits of science*. Cambridge: Cambridge University Press.

Dole, S. H. (1964). *Habitable planets for man*. New York: Blaisdell.

Fodor, J. A. (1975). *The language of thought*. Cambridge, MA: Harvard University Press.

Freudenthal, Hans. (1985). LINCOS: Design of a language for cosmic intercourse. In Edward Regis Jr. (ed.), *Extraterrestrials: Science and alien intelligence*. Cambridge: Cambridge University Press (pp. 215–228).

Goodman, N. (1954). *Fact, fiction and forecast*. Cambridge, MA: Harvard University Press.

Hart, M. H. (1982). Atmospheric evolution, the Drake equation, and DNA: Sparse life in an infinite universe, in M. H. Hart & B. Zuckerman (eds.), *Extraterrestrials—Where are they?* New York: Pergamon Press, pp. 154–165.

Jones, H. S. (1940). *Life on other worlds*. New York: Mentor.

Kukla, A. (1991). Criteria of Rationality and the Problem of Logical Sloth, *Philosophy of Science 58*, 486–490.

Kukla, A. (1992). Endogenous constraints on inductive reasoning. *Philosophical Psychology, 5*, 411–425.

Kukla, A. (2000). *Social constructivism and the philosophy of science*.

Kukla, A. (2005). *Ineffability and philosophy*. London: Routledge.

Latour, B., & Woolgar, S. (1979). *Laboratory life: The social construction of scientific facts*. London: Sage.

Laurence, S., & Margolis, E. (2001). The poverty of the stimulus argument. *British Journal for the Philosophy of Science, 52*, 217–276.

Lucretius. (1951). *On the nature of things*. London: Penguin Books.

Maratsos, M. p. (1989). Innateness and plasticity in language acquisition. In M. L. Rice & R. L. Schiefelbusch (eds.), *The teachability of language* (pp. 105–125). Baltimore: Paul H. Brookes.

Mash, R. (1993). Big numbers and induction in the case for extraterrestrIntelligence, *Philosophy of Science 60*, 204–222.

Matthews, R. (2001). Cowie's anti-nativism. *Mind and Language, 16*, 205–230.

Mayr, E. (1985). The probability of extraterrestrial intelligent life, in E. Regis Jr. (ed.), *Extraterrestrials: Science and alien intelligence*. Cambridge: Cambridge University Press, pp. 23–30.

McConnell, B. (2001). *Beyond contact: A guide to SETI and communicating with alien civilizations*. Beijing: O'Reilly.

McDonough, T. R. (1987). *The search for extraterrestrial intelligence: Listening for life in the cosmos*. New York: Wiley.

McMullen, E. (1980). Persons in the Universe, *Zygon 15,* 69–89.

Paine, T. (1948). *The age of reason*. New York: Liberal Arts Press.

Pinker, S. (1995). Facts about human language relevant to its evolution. In J.-P. Changeux & J. Chavaillon (eds.), *Origins of the human brain* (pp. 262–285). New York: Oxford University Press.

Putnam, H. (1967/1980). The innateness hypothesis and explanatory models in linguistics. In H. Morick (ed.), *Challenges to empiricism* (pp. 240–250). Indianapolis: Hackett.

Raup, David M. (1985). ETI without intelligence. In Edward Regis Jr. (ed.), *Extraterrestrials: Science and alien intelligence* (pp. 31–42). Cambridge: Cambridge University Press.

Regis, E. Jr. (1985). SETI Debunked, in E. Regis Jr. (ed.), *Extraterrestrials: Science and alien intelligence*. Cambridge: Cambridge University Press, pp. 231–244.

Rescher, N. (1985). Extraterrestrial science. In Edward Regis Jr. (ed.), *Extraterrestrials: Science and alien intelligence*. Cambridge: Cambridge University Press (pp. 83–116).

Rescher, N. (1998). *Complexity: A philosophical overview*. New Brunswick, NJ: Transaction Publishers

Rood, R. T., & Trefil, S. J. (1981). *Are we alone? The possibility of extraterrestrial civilizations*. New York: Scribners.

Sagan, C. (1973). *The cosmic connection: An extraterrestrial perspective*. New York: Doubleday.

Sagan, C. (1980). *Cosmos*. New York: Random House.

Sagan, C. (1982). The search for who we are. *Discover 3,* 31–33.

Sagan, C. (1983). We Are Nothing Special. *Discover 4,* 30–36.

Salmon, W. C. (1966). *The foundations of scientific inference*. Pittsburgh: University of Pittsburgh Press.

Schick, T. W. (1987). Rorty and Davidson on alternate conceptual schemes. *Journal of Speculative Philosophy,* 1, 291–303.

Shklovskii, I. S., & Sagan, C. (1966). *Intelligent life in the universe*. San Francisco: Holden Day.

Tipler, F. J. (1985). Extraterrestrial intelligent beings do not exist, in E. Regis Jr. (ed.), *Extraterrestrials: Science and alien intelligence*. Cambridge: Cambridge University Press, pp. 133–150.

van Fraassen, B. C. (1980). *The scientific image*. Oxford: Oxford University Press.

van Fraassen, B. (1989). *Laws and symmetries*. Oxford: Clarendon Press.

von Mises, R. (1957) *Probability, statistics, and truth*. New York: Macmillan.

Wittgenstein, L. (1953). *Philosophical investigations*. Oxford: Blackwell.

Index

www.ingramcontent.com/pod-product-compliance
Lightning Source LLC
Chambersburg PA
CBHW021117210326
41598CB00017B/1480